an electronic companion to
molecular
cell biology™

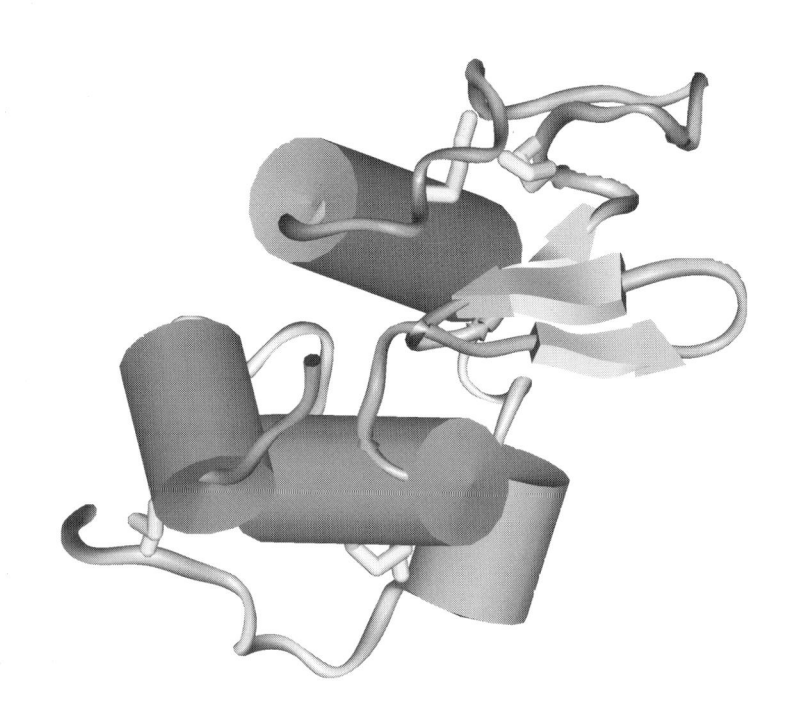

an electronic companion to
molecular
cell biology™

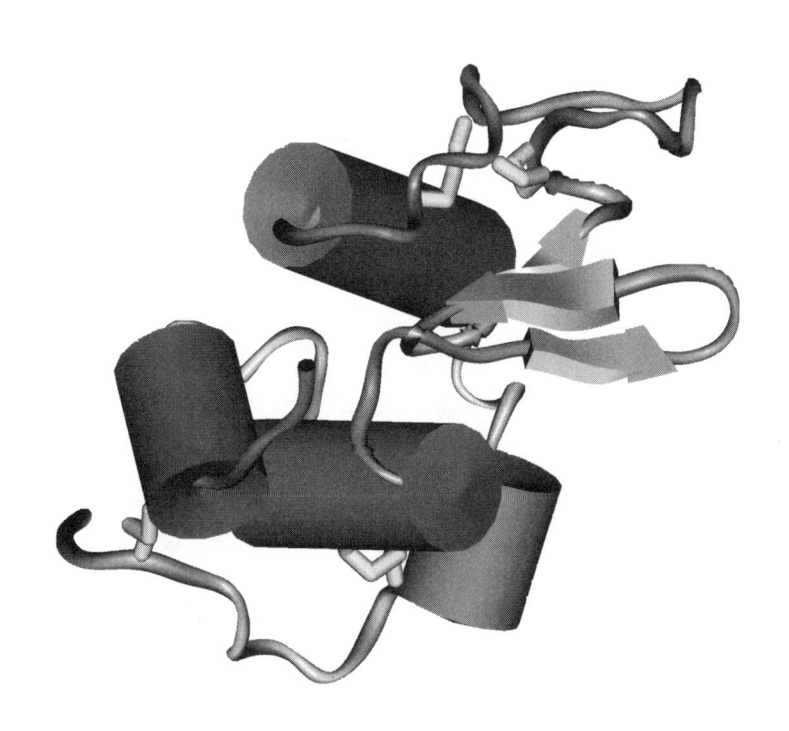

an electronic companion to
molecular cell biology™

Robert Van Buskirk
State University of New York, Binghamton

Udaya K. Liyanage

COGITO

Cogito Learning Media, Inc.
New York   San Francisco

No part of this book may be reproduced by any mechanical, photographic, or electronic process, or in the form of a phonographic recording, nor may it be stored in a retrieval system, transmitted, or otherwise copied for public or private use, without written permission from the Publisher.

© 1997, Cogito Learning Media, Inc.

ISBN: 1-888902-60-4

COGITO END-USER AGREEMENT

PLEASE READ THE FOLLOWING CAREFULLY BEFORE USING THIS PRODUCT (INCLUDING THIS WORKBOOK AND THE ACCOMPANYING SOFTWARE AND OTHER MATERIALS) (THE "PRODUCT"). BY USING THE PRODUCT, YOU ARE AGREEING TO ACCEPT THE TERMS AND CONDITIONS OF THIS AGREEMENT. IF YOU DO NOT ACCEPT THOSE TERMS AND CONDITIONS, PLEASE RETURN THE PRODUCT (INCLUDING ANY COPIES) TO THE PLACE OF PURCHASE WITHIN 15 DAYS OF PURCHASE FOR A FULL REFUND.

LIMITATIONS ON USE

Cogito Learning Media, Inc. ("Cogito") licenses the Product to you only for use on a single computer with a single CPU. You may use the Product only on a stand-alone basis, such that the Product and the user interface and functions of the Product are accessible only to a person physically present at the location of the computer on which the Product is loaded. You may not lease the Product, display or perform the Product publicly, or allow the Product to be accessed remotely or transmitted through any network or communication link. You own the CD-ROM or other media on which the Product is recorded, but Cogito and its licensors retain all title to and ownership of the Product and reserve all rights not expressly granted to you.

You may not copy all or any portion of the Product, except as an essential step in the use of the Product as expressly authorized in the documentation included in the Product. You may transfer your license to the Product, provided that (a) you transfer all portions of the Product (including any copies) and (b) the transferee reads and agrees to be bound by the terms and conditions of this agreement. Except to the extent expressly permitted by the laws of the jurisdiction where you are located, you may not decompile, disassemble or otherwise reverse engineer the Product.

LIMITED WARRANTY

Cogito warrants that, for a period of 90 days after purchase by you (or such other period as may be expressly required by applicable law) ("Warranty Period"), (a) the Product will provide substantially the functionality described in the documentation included in the Product, if operated as specified in that documentation, and (b) the CD-ROM or other media on which the Product is recorded will be free from defects in materials and workmanship. Your sole remedy, and Cogito's sole obligation, for breach of the foregoing warranties is for Cogito to provide you with a replacement copy of the Product or, at Cogito's option, for Cogito to refund the amount paid for the Product.

EXCEPT FOR THE FOREGOING, THE PRODUCT IS PROVIDED WITHOUT WARRANTIES OF ANY KIND, EXPRESS OR IMPLIED, INCLUDING, BUT NOT LIMITED TO, ANY IMPLIED WARRANTIES OF MERCHANTABILITY OR FITNESS FOR A PARTICULAR PURPOSE. Among other things, Cogito does not warrant that the functions contained in the Product will meet your requirements, or that operation of the Product or information contained in the product will be error-free. This agreement gives you specific legal rights, and you may also have other rights which vary from state to state. Some states do not allow the exclusion of implied warranties, so the above exclusion may not apply to you. Any implied warranties will be limited to the Warranty Period, except that some states do not allow limitations on how long an implied warranty lasts, so this limitation may not apply to you.

LIMITATION OF LIABILITY

In no event will Cogito be liable for any indirect, incidental, special or consequential damages or for any lost profits, lost savings, lost revenues or lost data arising from or relating to the Product, even if Cogito has been advised of the possibility of such damages. In no event will Cogito's liability to you or any other person exceed the amount paid by you for the Product regardless of the form of the claim. Some states do not allow the exclusion or limitation of incidental or consequential damages, so the above limitation or exclusion may not apply to you. Also, this limitation will not apply to liability for death or personal injury to the extent applicable law prohibits such limitation.

GENERAL

This agreement is governed by the laws of the State of California without reference to its choice-of-law rules. This agreement is the entire agreement between you and Cogito and supersedes any other understanding or agreements. If any provision of this agreement is deemed invalid or unenforceable by any court or government agency having jurisdiction, that particular provision will be deemed modified to the extent necessary to make the provision valid and enforceable, and the remaining provisions will remain in full force and effect. Should you have any questions regarding the Product or this agreement, please contact Cogito at 1-800-WE-THINK.

1 2 3 4 5 6 7 8 9 —RRD— 00 99 98 97 Printed in U.S.A.

contents

1   Cellular Chemistry 2

2   Protein Structure and Function 16

3   Nucleic Acids and Protein Synthesis 28

4   Techniques 40

5   Cell Membranes 54

6   Protein and Vesicular Traffic 68

7   Receptors and Second Messengers 80

8   Energy, Mitochondria, and Chloroplasts 92

9   Cytoskeleton and Cell Movement 110

10   Nervous and Immune Systems 128

11   Extracellular Matrix and Cell-Cell Interactions 154

12   The Cell Cycle and DNA Repair 172

13   Cancer and Cell Death 186

14   Development and Differentiation 202

15   Gene Expression in Eukaryotes and Prokaryotes 216

    Answers to Self-Testing Questions 227

    Glossary 295

an electronic companion to
molecular cell biology™

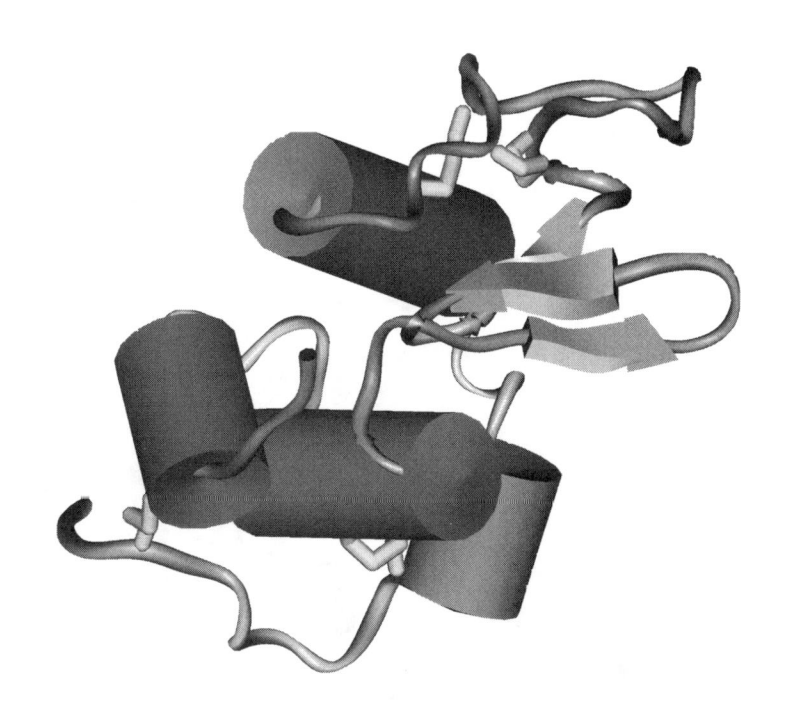

ATP

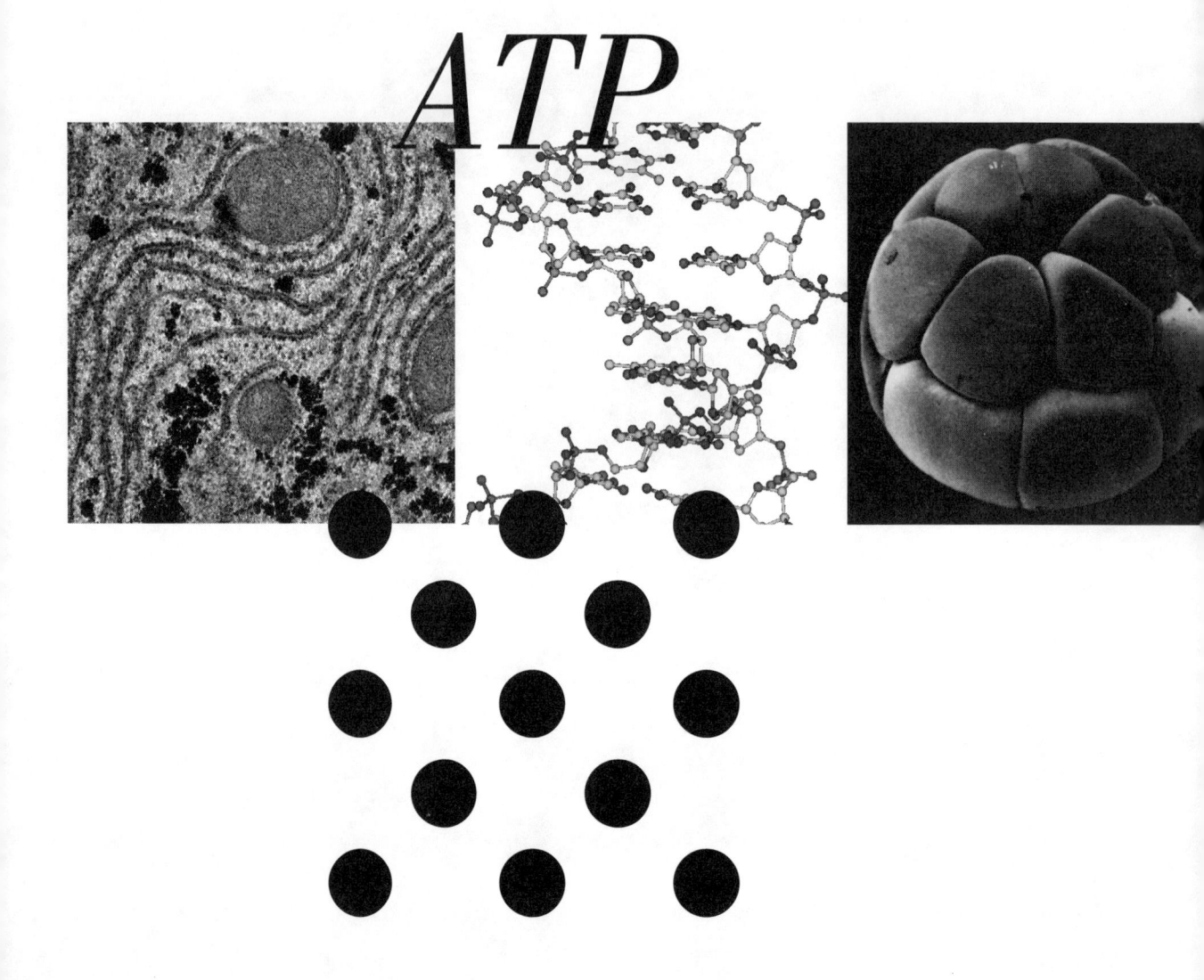

topic 1
Cellular Chemistry

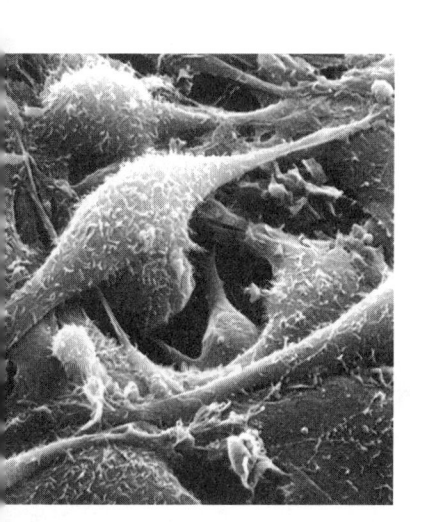

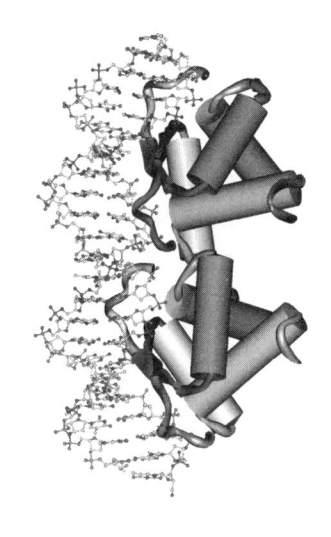

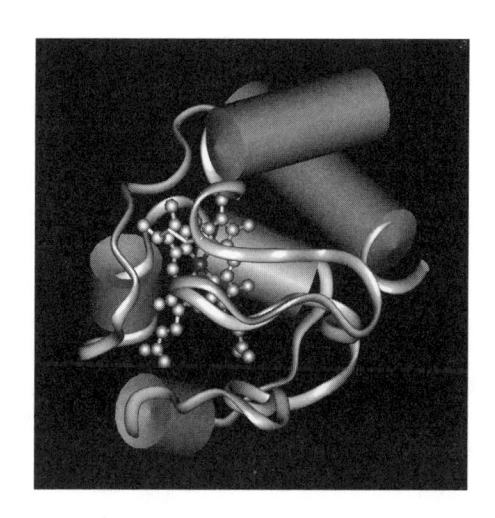

Summary

Water can be considered the solvent of life. Water serves as the medium in which most biochemical reactions occur in living cells. It is an excellent solvent for many **organic** and inorganic molecules. As a result, water is an ideal medium for biomolecules to meet and react with each other.

Both the angular shape of the molecule and the large difference in **electronegativity** between oxygen and hydrogen make water a **polar molecule**. Because of its higher electronegativity, oxygen attracts electrons more strongly than hydrogen does. This causes a small charge separation between hydrogen and oxygen, resulting in an electrical **dipole**. This dipole leads to the interactions known as **hydrogen bonds** that occur between two water molecules.

The hydrogen bond is a noncovalent interaction between two polar molecules, one of which has an unshielded positive charge on a hydrogen atom while the other has a negative charge on a highly **electronegative** atom like oxygen or nitrogen. The negative **free energy** associated with hydrogen bonding makes water a liquid rather than a gas at room temperature. This very important feature makes life possible on earth. If water were a gas at room temperature, it would not be a good solvent for biochemical reactions.

Hydrogen bonding also causes volume expansion during the freezing of water. As a result, ice has a lower density than water, allowing it to float on water. The ecological implications of this property are profound. Floating ice on ponds, lakes, and rivers preserves the thermal energy of the water underneath, protecting aquatic life through winter.

The dipolar property of water allows it to interact with other polar molecules. Ions such as sodium and chloride can therefore dissolve well in water. Conversely, nonpolar molecules like hydrocarbons that carry no charged groups and that are unable to make hydrogen bonds dissolve poorly in water. Such compounds separate into a nonaqueous phase when in water.

Carbon is a central atom in biomolecules. Because most of the chemical compounds in living organisms contain carbon, it was once erroneously thought that carbon-containing molecules were unique to life and were thus termed "organic." Carbohydrates, proteins, and lipids—all carbon containing molecules—make up the main structural and functional components of the cells. The nucleic acids that carry the cell's genetic information are polymers of carbon containing **nucleotides**. Plants store the solar energy harnessed in photosynthesis in the form of reduced organic compounds.

A carbon atom has four electrons in its outermost **orbital**; these can be shared with other atoms to make **covalent bonds**. Such bonds are often made with other common elements in cells like H, N, P, O, and S. Covalently linked chains of carbon can form the skeletons of large organic molecules like lipids. Two carbon atoms can also share two or three pairs of electrons between them to form double and triple bonds, respectively. Double bonds are rigid and planar, as they restrict rotation of each atom participating in the bond. This rigidity allows cis- and trans- isomers of the same molecule. When one carbon has four dissimilar groups attached, it is called a **chiral carbon.**

Covalently linked carbon chains can complete a circle to make ringed structures. Such ringed carbon skeletons are found in **amino acids** such as histidine. Alternate double bonds in benzene results in the movement of electrons that participate in the double bonds, making the **resonance** structure known as an *aromatic ring*. The amino acids phenylalanine and tyrosine contain aromatic rings. Multiple carbon rings can then join together to make complex molecules like cholesterol. Carbon rings that also contain nitrogen or oxygen are components of other biologically important molecules, such as nucleotides and carbohydrates.

Many amino acids are found in nature; however, only twenty alpha amino acids are incorporated into proteins. These can be represented in a generic structure of which the central atom is a carbon, known as the *alpha carbon*. Amino acids differ from each other by the content of the R group attached to this alpha carbon. Except in the case of glycine, where the R group is simply a hydrogen, all the other amino acids have a chiral alpha carbon and can therefore form mirror images of each other called *D* and *L enantiomers*. Only the L enantiomers of amino acids are found in proteins.

Amino acids are joined chemically to one another through a covalent bond between the carboxyl carbon of one amino acid and the amino nitrogen of another. This bond is referred to as a **peptide bond**. Due to resonance, peptide bonds have a partial **double bond** characteristic which makes them rigid. When two amino acids are combined by a peptide bond, the resulting molecule is called a *dipeptide*. Several amino acids can be similarly joined with peptide bonds to make a *polypeptide*. The term **protein** is reserved for one or more chains of polypeptides that together form a three dimensional structure. For example, insulin is a protein that consists of two polypeptides.

Once a polypeptide is formed, the amino acids in the molecule arrange in space to form one of many possible secondary structures. For example, a peptide chain can form a spiral coil known as an **alpha helix**. The alpha helix is a uniform structure with 3.6 amino acids occupying each turn of the helix. It is stabilized by the hydrogen bonds formed between the **alpha carboxyl groups** and the **alpha amino groups** of the amino acids in each successive turn of the helix. These hydrogen bonds can be broken and re-formed as the helix is stretched, rendering some elasticity to the protein. The alpha helix is the basic structural unit of the fibrous protein keratin that is found in hair, skin, nails, and wool.

Another common secondary structure is a flat, sheet-like arrangement known as the *beta pleated sheet*. Like the alpha helix, the beta pleated sheet is stabilized by hydrogen bonds, this time between the amino acids of adjacent peptide chains. As a result, the beta sheet is not as elastic as the alpha helix. The R groups of amino acids in a beta sheet are projecting out of the plane of the sheet. Depending on the directionality of the adjacent peptide strands, a beta pleated sheet can be antiparallel or parallel. Beta sheet structure is a recurring motif in many large proteins. Fibroin, the major component of silk, is characterized by a beta pleated sheet structure.

Rarely does a large protein consists only of alpha helices or beta pleated sheets. More often, different regions of a protein fold into different secondary structures that include alpha helices, beta pleated sheets, and other less ordered areas such as **random coils** and turns. This combination of secondary structures that determines the arrangement of amino acids is known as the **tertiary structure**. Tertiary structure is the highest level of organization observed in monomeric proteins consisting of only one polypeptide chain. However, some other proteins like immunoglobulins contain multiple polypeptide chains (**subunits**), each of which is necessary for function. The arrangement of these subunits in space is called the *quaternary structure* of the protein.

A protein's structure determines what functions it is most suited for. For example, the structure of hemoglobin allows it to bind oxygen efficiently in the lungs and then release oxygen in the tissues where it is needed. Enzymes are proteins designed to catalyze specific biochemical reactions. They are equipped with binding sites for their substrates and for the cofactors which enable them to function. Immunoglobulins are proteins which defend against invading pathogens. Other proteins such as actin serve as structural proteins that establish and maintain the cellular architecture. The degree of structural specialization of any protein bars it from performing the function of an unrelated protein. For example, **actin** cannot be a substitute for hemoglobin.

Two classes of nucleic acids are found in cells—*deoxyribonucleic acid* (DNA) and ribonucleic acid (RNA). Nucleotides, the basic subunit of both DNA and RNA, are covalently linked to make long strands of nucleic acids. A nucleotide is composed of three components:

- A five carbon sugar—deoxyribose in the case of DNA and ribose in the case of RNA.
- A phosphate group.
- A nitrogenous base.

Purine nucleotides contain the nitrogenous bases *adenine* (A) and *guanine* (G) while the pyrimidine nucleotides contain *cytosine* (C), *thymine* (T) and *uracil* (U). T is found only in DNA, while U is substituted for T in RNA. The term *nucleoside* refers to the purine or pyrimidine base bonded to the sugar. Nucleotides are formed by attaching the phosphates to the 5′ carbon atom of the sugar. This phosphate participates in a phosphodiester bond to link one nucleotide to another.

The amino acid sequences of all the proteins in a cell are determined by the genetic information encoded in DNA.

DNA is composed of two nucleotide strands entwined around each other to form a double helix, with ladder-like rungs between the two "backbones." The backbones consist of alternating sugar and phosphate groups while the crossrungs are composed of paired nitrogenous bases. The inherent spacing and polarity of the atoms within each nitrogenous base determines which other base it can form hydrogen bonds with. G binds exclusively with C while A binds exclusively with T (or U in RNA). The hydrogen bonds between these bases hold the two DNA strands together.

RNA is an integral part of the decoding machinery that forges amino acids into polypeptides. There are three major types of RNA—**messenger RNA (mRNA)**, transfer RNA (tRNA), and ribosomal RNA (rRNA)—all of which function in the process of protein synthesis.

Carbohydrates are composed of carbon, hydrogen, and oxygen. Sugars, starches and celluloses are the most abundant carbohydrates found in cells. Sugars serve as fuel for cells while starches are often storage molecules. Plant cell walls are made of celluloses.

Simple sugars like **glucose** and fructose are called *monosaccharides*. They conform to the general formula $(CH_2O)_n$. Most biologically important sugars have five or six carbons, multiple hydroxyl groups, and ketone or **aldehyde** as **functional groups**. Different isomers of sugars, resulting from the chiral carbons in the structure, can be distinguished by cellular enzymes.

Two monosaccharides can covalently link together through a C-O-C bridge called a **glycosidic bond** to make a *disaccharide* (e.g., **sucrose**). Polysaccharides are formed by linking many sugar molecules through glycosidic bonds (e.g., cellulose, a polymer of glucose). Unlike proteins and nucleic acids, polysaccharides can have branched chains. For example, **glycogen**, the major storage form of glucose in liver cells, has extensive branching that occurs at every 10–18 **residue** intervals. Polysaccharides are a major component of our food. The glycosidic linkages in most of them are hydrolyzed in the digestive system to yield monosaccharides. However, not all glycosidic linkages are amenable to hydrolysis by human **enzymes**. Therefore, not all carbohydrates can serve as food.

Lipids, like carbohydrates, are a heterogeneous group of compounds made of carbon, hydrogen, and oxygen, but they contain significantly less oxygen than do carbohydrates. As a result, they are much less **hydrophilic**. The biologically important lipids include neutral fats, phospholipids, steroids, and carotinoids.

The most abundant lipid in life is neutral fat, which is used by many organisms for long-term energy storage. Fat is composed of **glycerol** esterified with up to three molecules of **fatty acids**. Fatty acids that contain the maximum possible number of hydrogens are said to be saturated while the unsaturated ones contain some double-bonded carbon atoms. The level of saturation in fatty acids has important nutritional and medical consequences.

Phospholipids consist of fatty acids, glycerol, and phosphate. The most common phospholipids are composed of a glycerol molecule esterified with two fatty acids and a phosphate group which has an organic base such as choline attached to it. The phosphate end of a **phospholipid** molecule is polar, and therefore, soluble in water. The other end containing the hydrocarbon tails is nonpolar and insoluble. This **amphipathic** property makes phospholipids well suited for their role in cellular membranes.

Steroids, unlike other lipids, contain four interlocking rings of carbon. The nature of the side chains attached to these rings determines the type of the steroid. Among the most biologically significant steroids are **cholesterol**, hormones, and **bile acids**. Cholesterol, made only by animal cells, is an essential component of every animal cell membrane. It also serves as the precursor for the synthesis of adrenal cortical and sex hormones. Hormones regulate many aspects of metabolism in a variety of animals, from insects to humans. Bile salts, secreted by the liver into bile, **emulsify** fats in the intestines so that digestive enzymes can act on them.

Self-Testing Questions

1. Refer to the molecules in Figure 1.1 in answering the following questions.
   (a) Which of the molecules are able to make hydrogen bonds with water and dissolve?
   (b) Which compounds will separate into a nonaqueous phase when mixed with water?

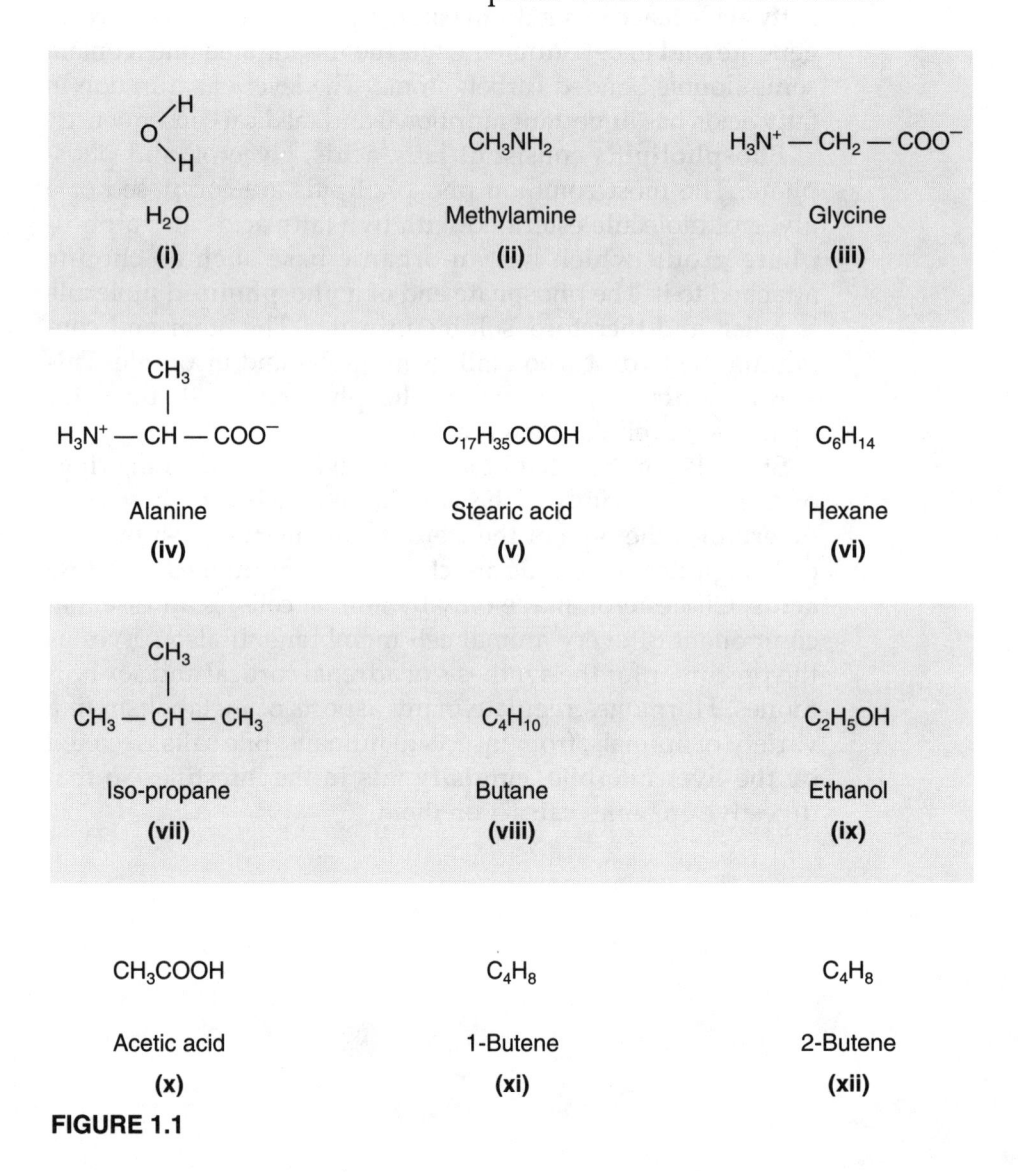

H_2O
(i)

CH_3NH_2
Methylamine
(ii)

$H_3N^+ - CH_2 - COO^-$
Glycine
(iii)

$H_3N^+ - CH(CH_3) - COO^-$
Alanine
(iv)

$C_{17}H_{35}COOH$
Stearic acid
(v)

C_6H_{14}
Hexane
(vi)

$CH_3 - CH(CH_3) - CH_3$
Iso-propane
(vii)

C_4H_{10}
Butane
(viii)

C_2H_5OH
Ethanol
(ix)

CH_3COOH
Acetic acid
(x)

C_4H_8
1-Butene
(xi)

C_4H_8
2-Butene
(xii)

FIGURE 1.1

(c) Which molecules contain chiral carbon atoms?
(d) Do any of the molecules have cis- and trans- isomers?
(e) Identify the molecules that have a permanent dipole moment and draw the site of the dipole. Compare the molecules you identified with the ones you chose as the answer for part (a). Are they the same or different?
(f) Which of the two molecules are structural isomers of each other?

2. **Van der Waal's** forces often determine the physical state of nonpolar organic molecules at room temperature. Arrange the molecules in Figure 1.2 in the ascending order of their melting temperatures. Give your reasoning.

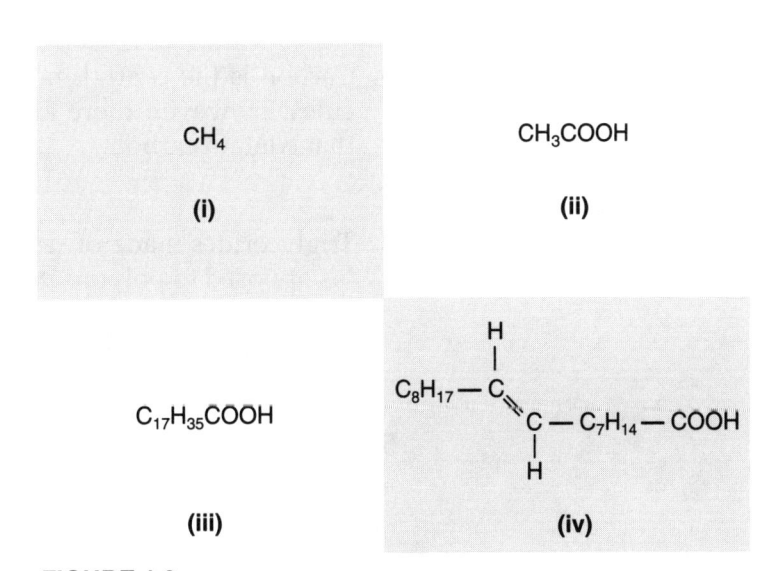

CH_4

(i)

CH_3COOH

(ii)

$C_{17}H_{35}COOH$

(iii)

$$C_8H_{17}-C\begin{matrix}H\\|\\ \\|\\H\end{matrix}C-C_7H_{14}-COOH$$

(iv)

FIGURE 1.2

3. (a) Review the section on protein structure. What levels of structural organizations can be disturbed by a single amino acid substitution of a protein?

(b) Your thesis project involves the study of bacterial protein A. You have identified a point mutation that renders the protein A nonfunctional. In your search for the revertants of mutant A phenotype, you discover that a mutation in protein B suppresses the mutant A phenotype. Upon further study you discover that wild type proteins A and B interact *in vivo* and that this interaction is necessary for function. Mutant A and mutant B interact equally well with each other. However, mutant A does not interact with wild type B and wild type A does not interact with mutant B. Using your knowledge of protein tertiary structure, explain these findings.

Indicate whether the following statements are True or False.

4. Carbon is the central molecule in most of the biomolecules. However, there are many "inorganic" materials that contain carbon.

5. Triglycerides made of saturated fatty acids and glycerol (as opposed to polyunsaturated fatty acids and glycerol), are solidified in cold temperature because more efficient hydrogen bonding occurs between saturated fatty acids.

6. Isothermal expansion of water at 4°C makes ice less dense than water. This property plays an important role in the survival of plants and animals in lakes and ponds during the winter.

7. Cholesterol is an important component of all animal cell membranes. The fluidity of lipid bilayers is regulated by the presence of cholesterol.

8. Ringed carbon containing molecules like benzene can be represented in two molecular structures. However, the real structure is a state somewhere in between these two structures. This is often referred to as resonance.

9. Fructose is a disaccharide found in fruit.

10. Lecithin, also called phosphatidylcholine, is a carbohydrate which is found on cell membrane proteins.

11. Most biologically important sugars are aldehydes or ketones which contain five or six carbons and multiple hydroxyl groups.

12. Phospholipids contain four interlocking carbon rings.

13. Bile acids secreted by the liver contain the steroid nucleus which is found in many other compounds, such as cholesterol and hormones.

Fill in the blanks of the following sentences to make accurate statements.

14. When sucrose is hydrolyzed, _____ and _____ are formed.

15. Human liver stores glucose in the form of _____ .

16. The components that make up a phospholipid are fatty acids, _____ , _____ , and _____ .

17. The primary structure of a protein refers to the _____ .

18. The most common secondary structures found in proteins are _____ and _____ .

Topic 1 Cellular Chemistry

Match the term on the right with its corresponding partner on the left.

19. Linoleic acid

20. A monosaccharide

21. This compound found in plant cell walls is a polysaccharide which contains monosaccharides linked by β1-4 glycosidic bonds

22. Organic carbon compounds which contain only carbon and hydrogen; a component of petroleum

23. A nucleotide derivative which is used as the energy currency of cells

24. A component of the nucleic acid backbone

25. Molecules with the same molecular formula but different structures

26. A carbon atom with four different groups covalently attached to it

27. A major component of cooking oil; this also is the major storage form of lipids in animal adepocytes

28. Building blocks of DNA

A. Triglycerides

B. A polyunsaturated fatty acid

C. Asymmetric carbon

D. Phospholipids

E. Fructose

F. Cellulose

G. Phosphate

H. Hydrocarbon

I. **ATP**

J. Chitin

K. Gasoline

L. **Nicotinamide adenine dinucleotide (NAD)**

M. Isomers

N. Purine

O. Deoxyribonucleotides

P. Buckyballs

Q. None of the above

ATP

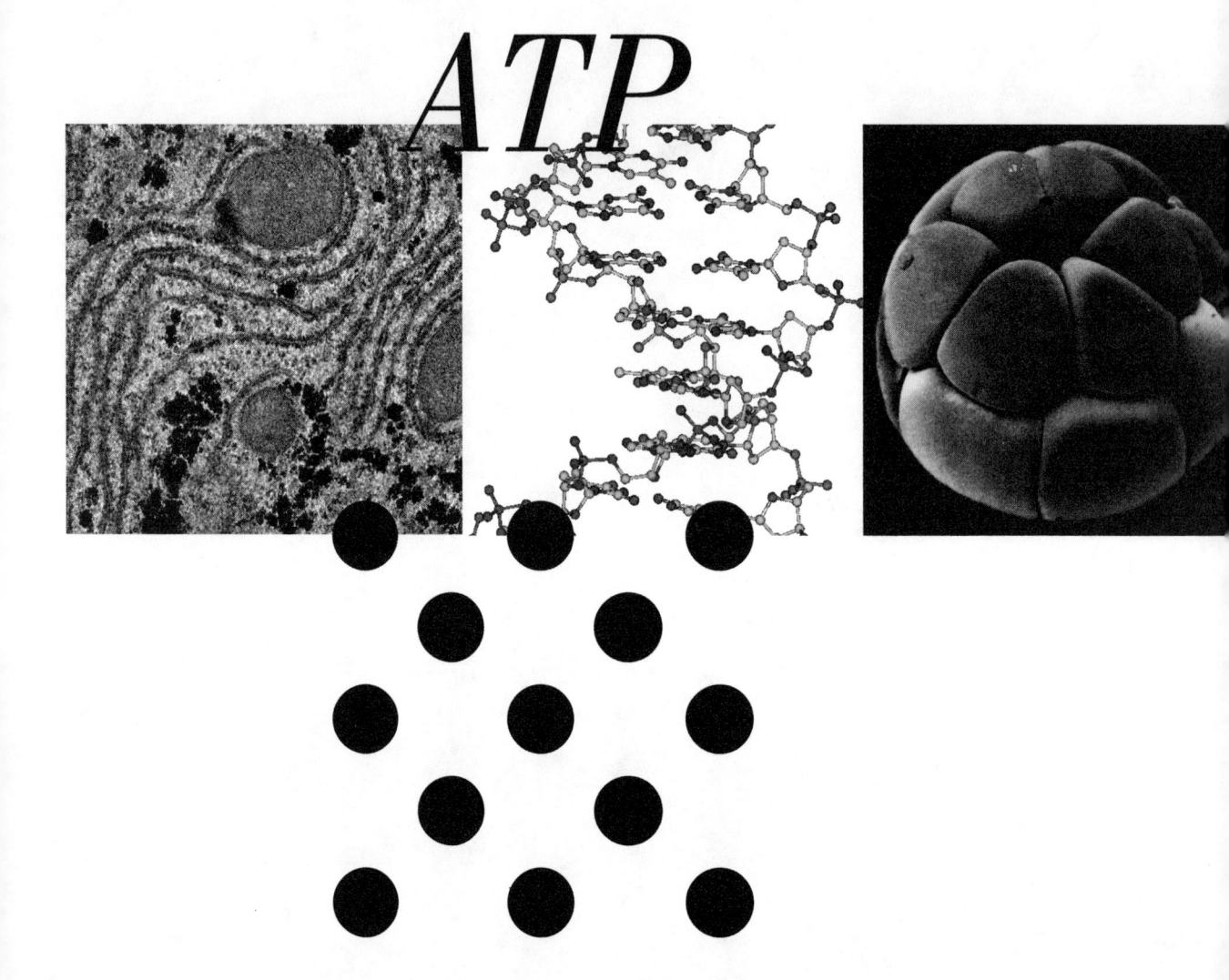

topic 2
Protein Structure and Function

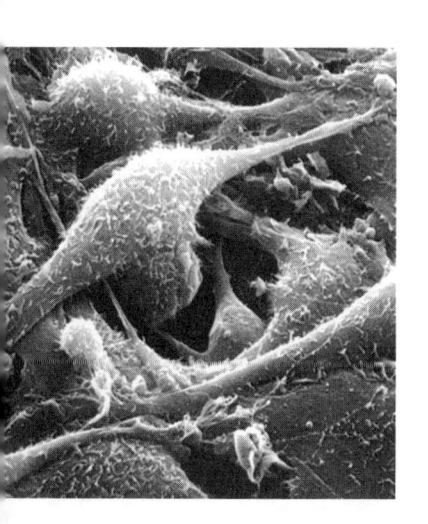

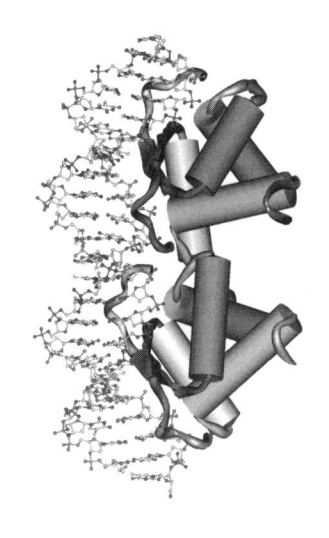

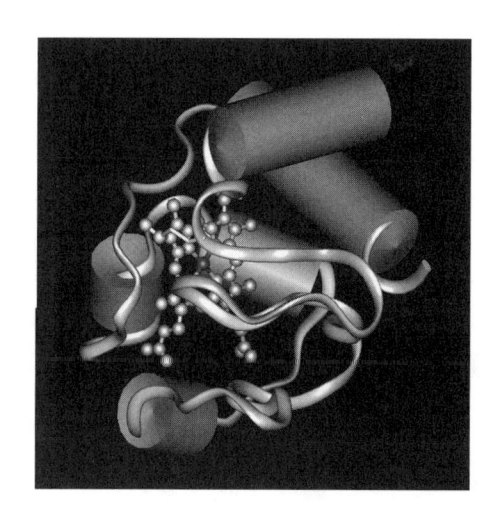

Summary

Proteins have crucial biological functions in cells. There are structural proteins that maintain cell shape, catalytic proteins that facilitate biochemical reactions, transport proteins that regulate material transport across the cell membrane or within the cell, storage proteins that convert critical biochemical constituents for long-term storage, and protective proteins to counteract environmental stresses. The integrity and proper functioning of the cell, and in turn of the organism, depend on proteins carrying out these tasks in an accurate and precisely regulated manner.

Proteins are polymers of smaller building blocks known as amino acids. Out of all the amino acids found in nature, only twenty are found in proteins. These amino acids have a typical structure that can be represented by a generic structural formula: all have a carboxyl group and an amino group attached to a central carbon atom known as the α carbon. The R group attached to the α carbon is what differentiates one amino acid from another. The chemical properties of the amino acid depend on the chemical properties of the R group.

Every protein in a cell usually carries out only one or a very limited number of tasks. As a result, there are thousands of proteins in a cell that perform specific functions meeting the diverse needs of the cell. The function of each protein is closely related to its structure. The structure is primarily determined by the amino acid sequence of the protein, which in turn is dictated by the genetic information in the gene that encodes the protein. The vast functional diversity of proteins is a result of the nearly unlimited ways that amino acids with different chemical properties can join to form proteins of different sizes, shapes, and biochemical characteristics.

Four levels of structural organization can be recognized in proteins. The first and most simple level is the linear order of amino acid residues linked together by covalent peptide bonds. This is known as the **primary structure** of a protein. The second level, known as the **secondary structure**, involves the folding of the polypeptide chain into ordered structures such as alpha helices and beta pleated sheets. The precise arrangement of secondary structures in the three dimensional space is referred to as the *tertiary structure*. The *quaternary* *structure* is the arrangement and stoichiometry of different polypeptides in a multimeric structure.

Protein structure is stabilized by both covalent and noncovalent interactions. Primary structure is established by the peptide bonds that form between the alpha amino and the alpha carboxylic groups of adjacent residues in the polypeptide chain. The higher order structural arrangements depend on **disulfide** bonds and noncovalent interactions like ionic bonds, **hydrophobic** and hydrophilic interactions, and hydrogen bonds. Disruption of these interactions by chemical agents such as urea and formamide leads to protein **denaturation**. When denatured, proteins lose the structural features that enable them to function. Some denatured proteins can fold back to their previous forms when denaturing agents are removed. However, many proteins need the help of other specialized proteins (known as **chaperones**) for proper folding. The chaperones are found in all compartments of the cell and play crucial roles in ensuring that cellular proteins are properly folded.

After being synthesized, many proteins are modified in numerous ways. These modifications are collectively known as *post-translational modifications* and are necessary to ensure the proper function of the protein. The most common post-translational modification is acetylation of the N-terminal amino acid residue. Over 80% of the proteins in a cell are acetylated. Acetylation is believed to stabilize proteins, as nonacetylated proteins are rapidly degraded in cells.

Often, segments of the protein are removed by proteolysis carried out by specific proteases. Certain proteins depend on such proteolytic cleavage to attain their final structure. Insulin, for example, is synthesized as a nonfunctional protein referred to as preproinsulin. Preproinsulin contains an N-terminal segment known as the signal sequence which carries the signal marking insulin as a protein to be secreted. In the endoplasmic reticulum and the **Golgi** complex, the signal sequence and a segment from the middle of the protein are cleaved off and the molecule is transformed into its biologically active form, insulin.

Some proteins are modified by attaching sugars to certain serine or asparagine residues. Such **glycosylations** play a role in the proper folding of some proteins, while in other proteins they function as molecular signals determining their intracellular location. For example, phosphorylated mannose residues (mannose-6-phosphate, or M6P groups) attached to lysosomal proteins function as **ligands** that bind to receptors located on lysosomal membranes. The receptor-ligand interaction between M6P and M6P receptors ensures that lysosomal proteins are concentrated in **lysosomes**.

Modification by the attachment of long chain fatty acids or glycosyl **phosphatidyl inositol** (GPI) results in the targeting of certain proteins to the plasma membrane. The hydrophobicity of the fatty acid or GPI molecule allows for insertion into the hydrophobic lipid bilayer of the cell membranes. These modifications essentially serve as membrane anchors for the protein.

Cells are dynamic entities with changing needs. As a result, the need for the activity of certain proteins changes with time. Certain activities needed at one moment may be harmful, disruptive, or simply unnecessary at another time. Therefore, cells regulate proteins by selective activation and inactivation. One of the strategies cells use for selective activation or inactivation is to attach phosphate groups to serine, threonine, or tyrosine residues. The added phosphate groups cause **conformational** changes in the protein, resulting in a newly active or inactive molecule, depending on the protein. For example, glycogen synthase, the enzyme that performs the rete limiting step in glycogen synthesis from glucose, is inactivated by phosphorylation. Glycogen phosphorylase, which catalyzes the physiologically opposite reaction to release glucose from glycogen, is activated by phosphorylation. In contrast, when glycogen is needed, dephosphorylation activates the final step in glycogen synthesis while also inactivating the enzymes responsible for glycogen breakdown.

Certain unwanted proteins are removed from the cell by proteolytic destruction. These proteins are targeted for destruction by tagging them with another protein called *ubiquitin*. Ubiquitin ligase is the enzyme that catalyzes the attachment of ubiquitin. Subsequent proteolysis of ubiquitinated proteins occur at the cytoplasmic bodies known as **proteosomes**.

The protein **cyclin** B is regulated by ubiquitin-mediated proteolysis. It is a component of the larger protein complex known as **maturation promotion factor (MPF)**, which is responsible for the mitotic cycling of cells. During interphase of the cell cycle, synthesis of cyclin B results in increased MPF activity, inducing the cell to enter mitosis. During mitosis, cyclin B is polyubiquinated and destroyed, allowing the cells to exit mitosis and enter interphase.

Self-Testing Questions

1. Which of the twenty amino acids found in proteins are likely to make up a membrane spanning helix? Explain your reasoning. In some multi-membrane spanning proteins, which are believed to make aqueous pores, there are amino acid residues bearing hydrophilic R groups. Speculate on the function of such residues.

2. Hemoglobin, the protein that carries oxygen in blood, is a multimeric protein which consists of two α subunits and two β subunits. One of your classmates has decided to study the differences between normal adult hemoglobin and sickle hemoglobin for her senior thesis project. She has managed to obtain both types of blood samples and separated hemoglobin from other blood proteins. Then she resuspended the hemoglobin in a solution containing a high amount of salt.
   (a) From your knowledge of protein structure, predict the effect of high salt on the structure of the protein. What level of protein structural organization is affected by salt? What level of structure is unaffected?
   (b) Your friend makes the solution acidic by adding hydrochloric acid. What would be the effect of the low pH on protein structure?
   (c) The hemoglobin in the solution is now subjected to digestion by a mixture of proteases in the presence of a reducing agent like b-mercaptoethanol. What do the proteases and the reducing agent do?
   (d) The resulting digest from part (c) is now analyzed by a two dimensional electrophoresis technique that separates the protein fragments according to their charge and size. When she analyzes the results, she notices that both normal adult hemoglobin and sickle hemoglobin give exactly the same pattern of peptide fragments, except for one sickle hemoglobin fragment which exhibits a slightly different electrophoretic mobility. Explain this result by using your knowledge of sickle hemoglobin.

3. Patients suffering from the severe genetic disease known as I-cell disease are found to have **fibroblasts** and **macrophages** containing lysosomes filled with various glycolipids and extracellular components. In these patients, the major functional abnormality is the inability to retain lysosomal enzymes responsible for digesting the accumulated glycolipids and cell debris. Investigations have revealed that these individuals are able to make functional lysosomal enzymes. However, instead of channeling them to lysosomes, these enzymes are secreted.

   (a) Using your knowledge on the role of protein modifications on targeting of proteins, suggest possible defects and mechanism(s) to explain the phenotype of patients with I-cell disease.

   (b) It was observed that cultured fibroblasts from patients with I-cell disease are able to take up lysosomal enzymes from normal individuals that were added to the culture medium. These cells are even able to take up the patient's own enzymes when they are modified *in vitro* by a phosphorylating enzyme. What modifications would you make to the hypotheses you formulated in part (a)?

   (c) Gaucher's disease is a similar lysosomal storage disease where one specific lysosomal enzyme is found to be nonfunctional. The fibroblasts from these patients are able to uptake added enzyme from the culture medium. What receptor must be present on the cell surface for this to occur? Can you come up with a strategy for treating patients with Gaucher's disease using the observation made on cultured cells?

4. **Eukaryotic** proteins are often produced in large scale in bacteria by introducing a plasmid carrying the eukaryotic gene into bacteria. Bacterial protein synthetic machinery is recruited to produce the eukaryotic protein. This

strategy allows for the large scale production of proteins which can be used for further study. One of the disadvantages of this strategy is that many eukaryotic proteins produced in this method tend not to retain the biological activity that they normally exhibit in eukaryotic cells.

(a) By using your knowledge of protein structure, folding, and post-translational modification, can you explain this observation?

(b) Presenillin-1, the protein believed to play a crucial role in the pathogenesis of early onset Alzheimer's disease, is one of the proteins recombinantly expressed in bacteria. When the protein made in bacteria is purified, it is observed that another protein tightly bound to presenillin is co-purified. When the protein mixture is incubated in a buffer containing ATP and Mg^{++}, the tight binding of this protein is relieved and presenillin is released. What general class might this other protein belong to?

(c) Disulfide bond formation is very inefficient in proteins expressed in prokaryotes. What biochemical reason accounts for this observation?

Indicate whether the following statements are True or False.

5. Peptide bonds are formed between two cysteins of the same polypeptide chain and are important in maintaining the primary structure.

6. Heat, extremes of pH, and reducing agents are able to destabilize the tertiary structure of a protein.

7. All enzymes are proteins.

8. Some proteins contain amino acids that carry covalently attached carbohydrate chains. These amino acids acquire the carbohydrates before they get incorporated into proteins.

9. Proteolytic cleavage of some proteins leads to biological activation of the protein.

10. Polypeptides shorter than 25 amino acid residues in length are rarely biologically active.

11. Acetylation of the N-terminal amino acid residue makes a protein more stable.

12. GPI anchors help localize a protein to the endoplasmic reticulum.

13. Mannose-6-phosphate receptors bind lysosomal proteins.

14. Protein kinases modify serine and threonine residues of a protein by phosphorylation. Such phosphorylations alter the tertiary structure of the protein.

Fill in the blanks of the following sentences to make accurate statements.

15. Proteins that consist of multiple subunits associated to perform a function are said to have a _____ structure.

16. The class of proteins that help proper folding of large proteins *in vivo* are called _____ .

17. Hydroxylated proline and glysine residues are common in the most abundant protein in life, _____ .

18. The site on the tertiary structure of an enzyme that can bind small molecules and alter the conformation at the catalytic site is known as the _____ site.

19. Cell shape is maintained by _____ proteins like actin that make up the skeletal frame of a cell known as the _____ .

20. Water soluble proteins have _____ residues at the core of the protein globule while the outer surface contains _____ residues.

21. Amino acids of alpha helices that span the membrane interact with the interior of the lipid bilayer. In general, the amino acids in these alpha helices have _____ R groups.

22. Proteases are protein enzymes that hydrolyze _____ bonds which link _____ of the _____ structure of the protein.

23. Glycine has the smallest _____ group of any amino acid. This property is important in allowing polypeptides to make tight turns as in the case of the alpha chain of collagen.

24. Independently folded regions called _____ are found in some large proteins. These small globular structures may or may not perform independent functions.

ATP

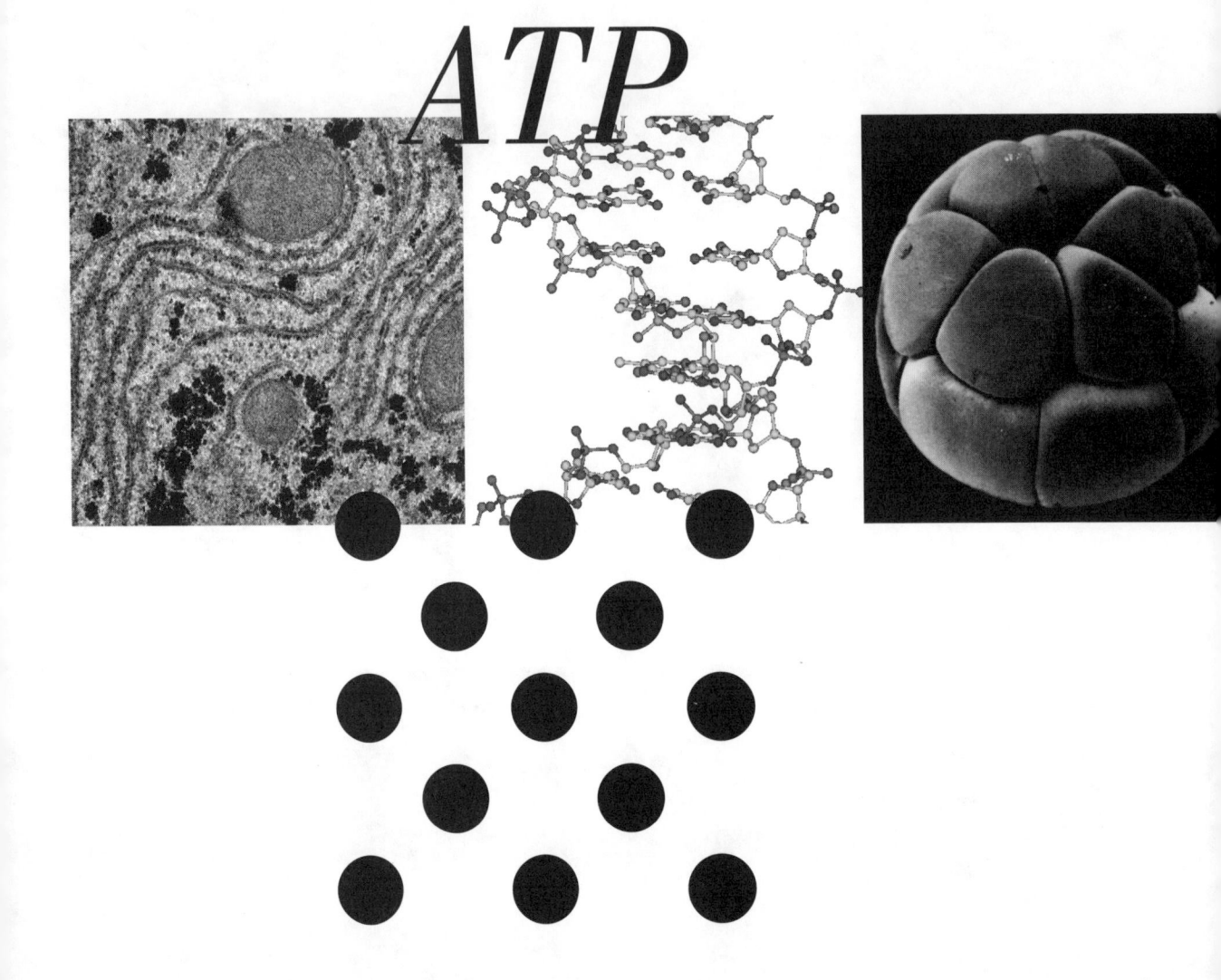

topic 3
Nucleic Acids and Protein Synthesis

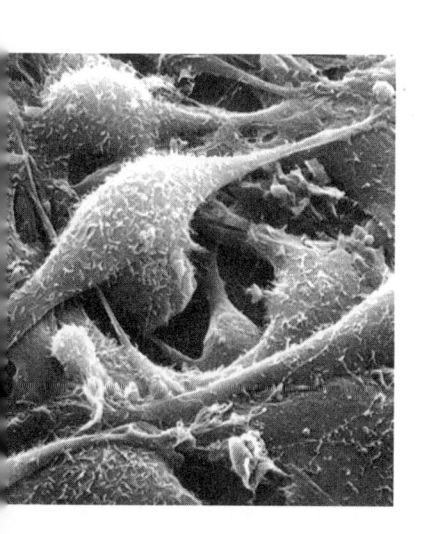

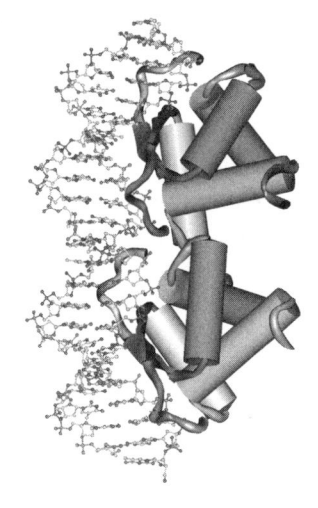

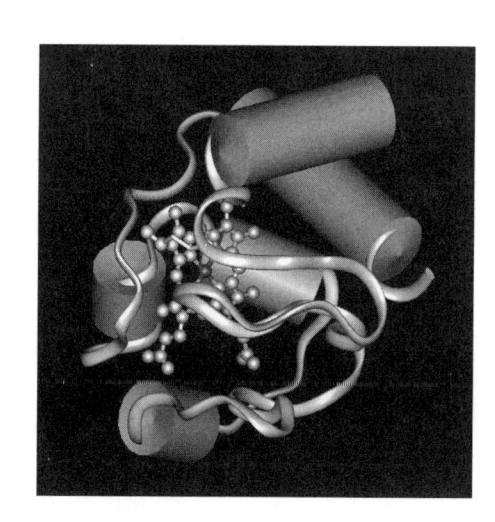

Summary

There are two types of nucleic acids. The first is deoxyribonucleic acid (DNA), which contains the information prescribing amino acid sequence for proteins. The other is ribonucleic acid (RNA), which helps forge amino acids into proteins.

DNA forms the genetic library containing the information necessary to build a cell and an organism. This information is stored in all living cells in a three lettered code (the **genetic code**) formed by the combination of four bases found in the nucleic acids. These bases are adenine (A), guanine (G), cytosine (C), and thymine (T) in DNA or uracil (U) instead of thymine in RNA. The information written in this code is transferred in the form of DNA from one generation to the next.

DNA is double stranded and has a right-handed helical structure. The two strands of a DNA molecule run in opposite directions (5' → 3'/3' → 5'), making the double helix antiparallel. In cells, the enormous amount of double stranded DNA is densely packed into chromosomes. First, DNA is wound around an octamer of **histones** to form nucleosomes. These nucleosomes then form a solenoidal array with six nucleosomes per turn. The solenoidal arrays are finally attached to a protein scaffold to make chromosomes.

The double stranded helical structure of DNA is mainly stabilized by hydrogen bonds between the complementary bases of the two DNA strands. Heat and chemical agents such as urea and formamide can disrupt these hydrogen bonds to separate the two strands (denaturation). When the denaturing agent is removed, the complementary strands may again find each other (renaturation). This property of DNA is exploited in many experimental techniques such as PCR and **Southern Blotting**. The temperature at which a particular molecule of double stranded DNA denatures is known as the melting temperature (T_m). Factors such as chemicals, salt, and the level of G-C content in a DNA molecule can affect the melting temperature.

Transmission of genetic information from the parent cells to daughter cells is ensured by a complete and almost error-free DNA replication process that produces a duplicate of the parent cells' genetic complement. DNA replication is accomplished by using the parent DNA as a template to synthesize two new strands. This process is said to be semi-conservative, as both daughter DNA molecules inherit one parent strand and one newly synthesized "daughter" strand. DNA replication is carried out by an enzyme known as **DNA polymerase**. In cells there are multiple such polymerases. Replication is often accompanied by a verification process that replaces incorrectly inserted bases with correct ones.

DNA replication is initiated by the annealing of a short piece of primer RNA to the template. The 3' OH group of this primer RNA carries out a **nucleophilic** attack on the α phosphate of the nucleotide triphosphate and splits off pyrophosphate. This process incorporates the first base of the new DNA strand. The newly incorporated base is complementary to the corresponding base on the template strand. The 3' OH of the new base then carries out the next nucleophilic attack and so on until the parent strand is fully copied. The DNA duplex is sequentially unwound as the **replication fork** progresses. As the DNA strand can grow in only one direction (5' $\rightarrow$ 3'), one continuous **leading strand** and many discontinuous **lagging strands** (**Okazaki fragments**) are formed. The Okazaki fragments are later ligated by an enzyme called DNA **ligase**. The RNA primers used in the initiation of replication are then also removed. Since the **genome** of a cell is quite large, replication is initiated at multiple points so that the whole genome is replicated quickly.

RNA is chemically similar to DNA except that the sugar in RNA is ribose (an -OH group is attached to the 2' carbon in ribose). This makes RNA more labile than DNA. RNA also incorporates the **pyrimidine** uracil rather than thymine. RNA is single stranded but can hybridize with DNA through base pairing. An RNA strand can also make many secondary structures with itself such as hairpins and stem loops. The tertiary structures of some RNAs (e.g., tRNA) are crucial for their function.

There are three main types of RNA, all copied (transcribed) from DNA templates. Messenger RNA (mRNA) determines the amino acid sequence of a protein. The majority of RNA in a cell is ribosomal RNA (rRNA) which, along with many proteins, makes the **ribosomes**, the structures on which amino acids are linked together to make proteins. Transfer RNA (tRNA) carries amino acids and helps insert them in correct sequence according to the genetic code on the mRNA. These three RNAs are the key elements of the protein synthetic machinery that decodes genetic information to synthesize proteins. However, some RNA also functions enzymatically to facilitate biochemical reactions.

Transcription is the process of copying DNA to produce the complementary sequence in RNA form. This process is **catalyzed** by the enzyme called RNA polymerase (RNAP). RNAP starts transcription at special initiation sites called promoters at the beginning of **genes**. Transcription initiation in prokaryotes is helped by the protein factor σ in *E. coli*. In eukaryotes, transcription initiation is a complex process with many proteins (**transcription factors**) joining in a stepwise manner to make the transcription initiation complex. RNAP starts transcription from defined sites known as start sites and continues until it meets a terminator signal at the end of the gene. Termination can be caused by the rapid formation of a stable hairpin structure in the newly synthesized RNA chain. The RNA polymerase pauses when it encounters such a hairpin, the RNA-DNA hybrid dissociates, and the RNAP releases the DNA. Transcription of other genes can be terminated with the help of termination factors such as ρ.

In eukaryotes, the RNA **transcripts** are processed before they become functional. For example, functional mRNA is made by **splicing** the functional portions of the sequence (**exons**) together and removing others (**introns**). A methyl guanosine cap is then added to the 5' end and a poly A tail to the 3' end to make functional mRNA. Ribosomal RNA is transcribed as a single large molecule which is cleaved by specific nucleases. These smaller rRNA molecules and ribosomal proteins together make subunits of ribosomes. Certain segments of tRNAs are also spliced off before they fold into very specific tertiary structures. Several of the bases in tRNA are modified and an amino acid is attached to the acceptor arm by a class of enzymes called **aminoacyl-tRNA** synthetases. Each tRNA type has its own aminoacyl-tRNA synthetase which catalyzes this reaction.

As mentioned before, genetic information is written in DNA and RNA in three-lettered words (**codons**) using an alphabet of four letters (A, G, C, and T or U). This allows 64 (4^3) unique codons. Out of these, 61 code for amino acids and three act as **translation** stop signals. Since there are only 20 amino acids found in proteins, they can be represented with more than one codon. This property of having more than one codon for each amino acid gives rise to the **degeneracy** of the genetic code—

the third letter of the codon may vary without changing the amino acid it signifies. The genetic code is generally "universal," i.e., it does not vary considerably among species.

Any change in the nucleotide sequence of a gene may result in a protein that has an altered amino acid sequence and even an altered function. For example, a single base substitution can change a codon to result in an amino acid substitution (a **missense mutation**) or a prematurely terminated protein (a **nonsense mutation**). A missense mutation may be "tolerated" if the structure of the protein is not significantly altered, or may alternatively be devastating if it alters the protein's function. Removal or addition of one or more bases (deletions and insertions) may result in a frame shift which causes the amino acid sequence to change completely. Such mutant sequences usually cause chain termination during protein synthesis due to new **stop codons** that result from the frame shift.

Translation of the code in an mRNA molecule into a protein involves multiple steps, enzymes, and **cofactors**. Ribosomes provide the catalytic surface on which mRNA and **activated tRNA** meet and interact. Translation occurs in three phases. Initiation involves the assembly of mRNA and a ribosome along with the first activated tRNA molecule and its associated amino acid to the P site of the ribosome. The **chain elongation** phase occurs as the α amino group of the amino acid at the A site carries out a nucleophilic attack on the α carboxyl group of the amino acid at the P site. Elongation continues as the ribosome moves along the mRNA codon by codon. The tRNA molecules associate with mRNA codons through their corresponding anticodons; thus, the sequence of mRNA codons determines the sequence of amino acids that form peptide chains. This mRNA-tRNA interaction is somewhat flexible with respect to the third base in the mRNA codon, an effect referred to as **wobble**.

Protein synthesis in bacteria begins when a **purine**-rich sequence of nucleotides near the start site of mRNA (the Shine-Dalgarno sequence) is identified through base pairing with the small ribosomal (16S) RNA. **Eukaryotic** ribosomes appear to identify the 5' cap instead. After the cap is located, the ribosomal subunit is thought to slide along the mRNA until it finds an AUG. This is often flanked by a sequence (the *Kozak sequence*) which favors translation initiation at the AUG. **Met-tRNA**, assisted by a protein-**GTP** complex, then binds to the mRNA near the AUG. A group of proteins called *initiation factors* help **small ribosomal subunits** find the mRNA. Once the **large ribosomal subunit** binds through the ribosomal binding sequence in mRNA, the initiator complex is complete.

Chain elongation starts with the binding of the second aminoacyl tRNA to the ribosomal A site. In bacteria, the proteins Tu and Ts help this binding. In eukaryotes, factors EF1 and $EF_{1\beta}$ do the same. An activated Tu-GTP complex binds to the TψCG loop on tRNA and mediates binding to the ribosome. Once the codon and the **anticodon** are properly paired, the peptide chain of the **peptidyl tRNA** is transferred to the newly arrived aminoacyl tRNA, making a peptide bond. Now the ribosome moves one codon down the mRNA, ejecting the tRNA at the A site and translocating the peptidyl tRNA to the P site. This reaction, which uses GTP, is catalyzed by EF_G in bacteria and EF_2 in eukaryotes.

The process of chain elongation continues until a **terminator codon** is reached. Once the ribosome arrives at the stop codon, translation is stopped with the help of **termination factors**. Three proteins serve as termination factors in bacteria, while there is only one known in eukaryotes. The process requires GTP hydrolysis. The polypeptide chain attached to the peptidyl tRNA is released, and the two ribosomal subunits are recycled for another translation cycle.

Many widely used **antibiotics** selectively kill bacteria by inhibiting bacterial protein synthesis machinery. They can disrupt various steps in the translation process. Their specificity of action against bacteria results from the differences in bacterial and eukaryotic ribosomes.

Self-Testing Questions

1. Partial sequence of the cDNA of a protein is given below. Only the sequences at the beginning and the end of the clone are given.

   5'-CCACCATGGGGAGTCT**C**AGCCAGAGC...
   TCCCCCGACTCCTGGGTGTGAGG-3'

(a) Design two primers that you would use for PCR of this cDNA clone. What are the desired characteristics of these primers?
(b) What other reagents need to be added to the PCR reaction to make it work?
(c) If you would like to have the DNA that you produce out of your PCR to be labeled with radioactive ^{32}P, which phosphate of the nucleotide triphosphate (NTP) would you label with the radioactive isotope?

2. (a) Refer to the DNA sequence given in question 1. Assuming that the first ATG you see toward the 5' end of the clone is the translation start site, predict which strand (i.e., the one given or the complementary strand) acts as the template for transcription.
(b) What is an open reading frame? Find the open reading frame of the sequence in question 1 and write the amino acids encoded. Can you find the stop codon at the end of the sequence?
(c) What will happen if an additional T is inserted after the C indicated by the boldfaced underlined character (an insertion **mutation**)?
(d) What is a Kozak sequence? Can you find the Kozak sequence in the DNA sequence given?
(e) What is a Shine-Dalgarno sequence? What is the function of such a sequence?
(f) Is there a Shine-Dalgarno sequence in that given in question 1?

Indicate whether the following statements are True or False.

3. The central dogma of molecular biology states that the information encrypted in DNA is copied to make RNA, which then directs the synthesis of protein molecules that carry on many cellular functions.

4. Nucleotides are made of three components: a sugar molecule, a phosphate group, and a glycerol molecule.

5. The primitive nature of bacterial genomes are revealed by large stretches of DNA that do not code for any proteins. This DNA is sometimes referred to as junk DNA.

6. DNA that contains more than 50% G and C has a higher melting temperature.

7. Small molecules of DNA can be reversibly denatured many times without permanent damage to the molecule.

8. In prokaryotes, RNA synthesis and protein synthesis are coupled processes. This is possible because there is no need of processing RNA to remove introns.

9. The codons in a DNA strand can be read in three different reading frames depending on which base the reading is started. In most of the genes, there is only one reading frame that is not interrupted by random stop codons in the middle. This is known as the open reading frame. In viruses like HIV, however, more than one reading frame is utilized for encoding proteins. This allows the viruses to use the same region of the genome to synthesize multiple proteins.

10. **Mitochondria** in eukaryotic cells contain circular DNA.

11. Certain antibiotics specifically inhibit bacterial protein synthesis because the bacterial genetic code is entirely different from the eukaryotic genetic code.

12. Certain genes contain a consensus sequence known as the Kozak sequence near the ATG initiator codon where translation starts.

Fill in the blanks of the following sentences to make accurate statements.

13. Nucleic acids are made of building blocks called _____ .

14. The building blocks in nucleic acids are linked together by _____ bonds.

15. The entire DNA sequence necessary for the synthesis of a protein is called a _____ .

16. In bacterial genomes, several genes devoted to one metabolic goal are clustered together so that coordinated expression of all such genes can be regulated. Such clusters of genes are known as _____ .

17. In Xeroderma Pigmentosum, a disease characterized by recurrent skin cancers, there is a deficiency in the enzyme _____ . This enzyme impairs the ability of a cell to repair the abnormal **thymidine**-thymidine dimers caused by UV radiation.

18. One of the features that distinguishes eukaryotic from **prokaryotic** genes is the presence of intervening sequences known as _____ in between the real coding segments of the gene, known as _____ .

19. The process of DNA replication is said to be _____ because each new daughter molecule of DNA inherits one of the parental strands.

20. The synthesis of a new DNA strand by DNA polymerase can progress only in the _____ direction. As the replication fork progresses, one DNA strand is synthesized as a continuous molecule while the other strand is synthesized in smaller fragments called _____ .

21. In eukaryotes, the processing of mRNA involves splicing the _____ together, adding a _____ to the 5' end, and adding a _____ to the 3' end.

22. In bacterial mRNA, a sequence known as the _____ sequence is located near the start site base pair with the 16S small ribosomal RNA. This sequence actively recruits ribosomes to mRNA to start protein synthesis.

ATP

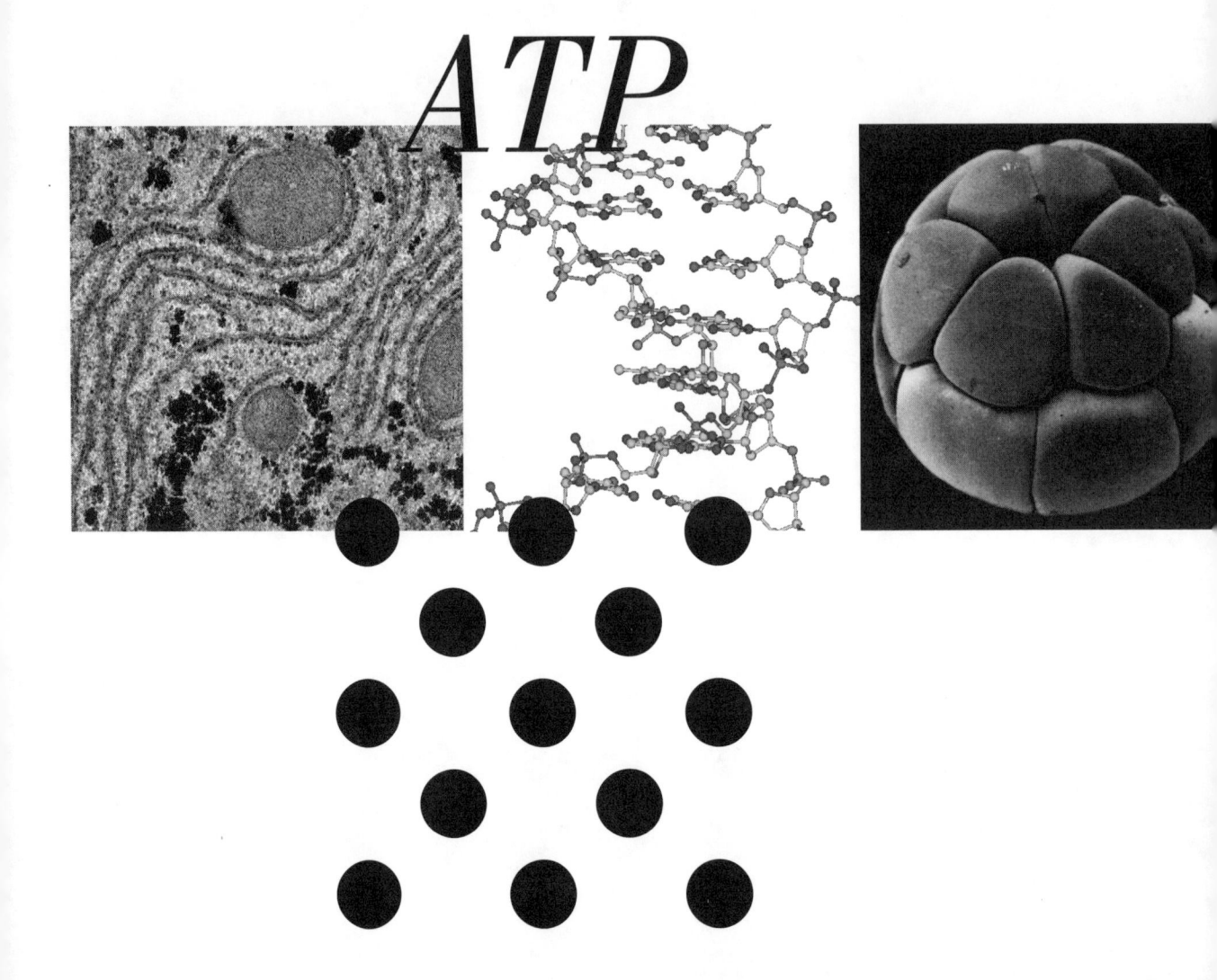

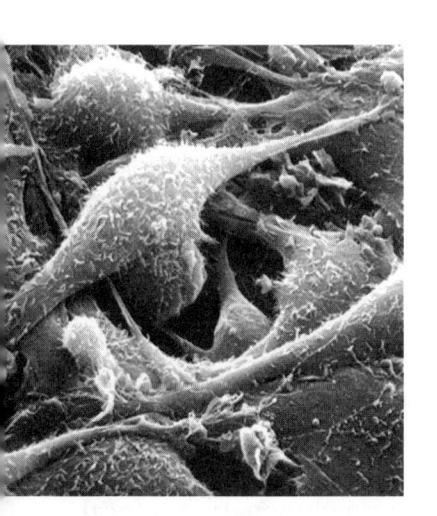

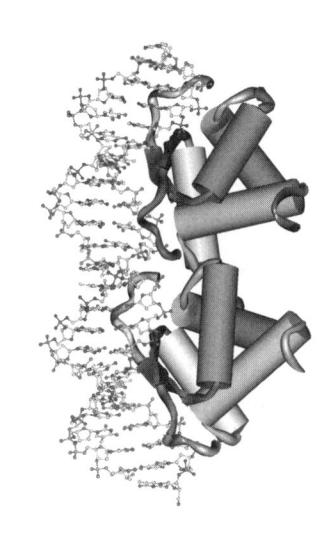

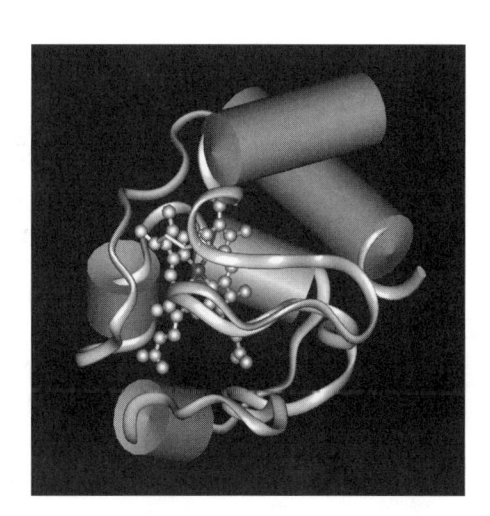

topic 4
Techniques

Summary

Our current understanding of how cells work is limited, in part, by the techniques and procedures used to study them. Many types of analytical equipment and procedures have been developed since the discovery of the light microscope that have helped cell biologists explore the cellular universe. While it is difficult to categorize all of them, it is possible to recognize several of the large groups of techniques such as:

- Light **microscopy**
- Electron microscopy
- Nucleic acid techniques
- Cell separation and fractionation
- Protein separation

Light microscopy was the original foundation for cell biology. The **bright-field microscope** was the first to be developed; it continues to be useful in the cell biology laboratory for examining fixed and stained cells. Other types of microscopy emerged that could manipulate the light path through cells so as to reveal different images. The **phase microscope** relies on the differences in refractive index in cells so that living cells can be viewed. The interference contrast microscope (Nomarski) can manipulate the light path so as to reveal the three-dimensional nature of the cell. **Fluorescence** microscopy became very popular in the 1970s and can be used to identify specific molecules in cells through fluorescence **immunocytochemistry**. Finally, in the 1980s the revolutionary **confocal microscope** was developed as a consequence of both computer and laser technology. The confocal microscope yields very crisp images and can optically section, by virtue of its ability to image the specimen point by point.

Light microscopy's cousin, electron microscopy, is governed by many of the same principles. For instance, in both transmission electron microscopy and light microscopy, resolution is dictated by **Abbe's equation**. The transmission electron microscope reveals internal structures of specimens with a resolution of single atoms. Like light microscopes, it has objective and condenser lenses, but it uses electrons as the illuminating source. The most commonly used procedure for the transmission electron microscope is the plastic thin sectioning technique, but internal membrane structure is best seen using an alternative procedure called *freeze-fracture*. The scanning electron microscope, in contrast, is designed to see cell surface structure and has a resolution considerably lower than that of the transmission electron microscope.

Nucleic acid techniques, like light microscopes, are quite varied. Many, however, are similar because they rely on the complementary nature of the DNA double helix. These hybridization techniques include Southern blots, a gel-blot procedure used to detect the presence and size of DNA sequences through the use of a complementary probe. Northern blots are used in a similar fashion to determine the presence, size, and amount of a specific mRNA species. Another procedure, DNA gel electrophoresis, is important in viewing restriction fragments cut from long, unwieldy pieces of DNA. The polymerase chain reaction also uses complementarity along with a heat-stable DNA polymerase to generate large amounts of DNA from only a single strand.

Proteins are studied by cell biologists more than any other type of molecule. This is, in part, because of their enzymatic qualities as well as their multifunctional applications in cells. Thus, **protein separation** techniques are the most commonly used molecular separation systems in the cell biology laboratory. **Ion-exchange chromatography** separates proteins according to charge attraction; **gel filtration** separates by physical size; affinity chromatography isolates proteins on the basis of the affinity of a ligand for a receptor; **SDS gel electrophoresis** separates proteins according to size and migration through a gel; while antibodies are commonly used in many different formats to separate proteins by exploiting an **antigen**-antibody relationship.

Many other techniques can't be grouped with the others listed here. They include **autoradiography**, a technique that can localize radioactive tagged molecules in a cell; and cell culture—a collection of techniques that allow the cell biologist to grow cells in an artificial environment.

Self-Testing Questions

Fill in the blanks of the following sentences to make accurate statements.

1. _____ equation is used to calculate resolution of light and electron microscope systems.

2. The _____ microscopes are designed to view living cells.

3. Antibodies are commonly used to localize molecules in cells by employing _____ microscopy.

4. A laser is a key component of the _____ microscope.

5. The interior of cell membranes can be viewed using _____.

6. Agarose gel electrophoresis is commonly associated with the separation of _____.

7. Northern blots are designed to detect _____ levels.

8. Amplification of genes is done through the technique of _____ , which uses *Taq* polymerase as a DNA polymerase.

9. **Reverse transcriptase** is used when doing _____ cloning.

Self-Testing Questions

10. SDS gel electrophoresis separates proteins by _____.

11. Two-dimensional gel electrophoresis separates proteins using two techniques: _____ and _____.

12. Immunoprecipitation uses _____ as the key tool to achieve separation of proteins in a solution.

13. Detection of radioisotopes in cells or in gels is called _____.

The following problems focus on normal human epidermal keratinocytes (NHEKs) used to construct an artificial, multi-layered human epidermis. Select the letter that best completes the following statements.

14. One of the problems with isolating NHEKs is that they are often contaminated with human dermal fibroblasts. These two cell types separate from each other in a centrifugal field. This characteristic could be exploited to separate the cells using _____.
 (a) fluorescence-activated cell sorting
 (b) velocity sedimentation
 (c) dynabeads
 (d) selective surfaces
 (e) panning

15. Once the NHEKs have stratified to form a complete epidermis, you decide to process the tissue with paraffin and stain it with hematoxylin and eosin. In so doing you are using _____.
 (a) transmission electron microscopy
 (b) two-dimensional gels
 (c) bright-field microscopy

(d) western blots
(e) phase microscopy

16. An increase in extracellular calcium is critical to the differentiation of NHEKs into stratified human epidermis. When you increase extracellular calcium, you want to simultaneously follow what happens to intracellular calcium in the NHEKs. To measure this, you would choose which of the following techniques: _____.

(a) PCR
(b) fluorescence microscopy
(c) freeze-fracture
(d) subtractive hybridization
(e) SDS gel electrophoresis

17. During the process of propagating NHEKs in culture, you want to observe them without having to fix and kill them. Which of the following techniques would allow you to do this? _____

(a) two-dimensional gel electrophoresis
(b) transmission electron microscopy
(c) scanning electron microscopy
(d) Nomarski microscopy
(e) freeze-fracture

18. NHEKs change their cytoskeleton (**actin**, tubulin, and keratins) quite dramatically when first presented with an increase in extracellular calcium. Which of the following would be best to view this in optical sections? _____

(a) polarizing light microscope
(b) phase microscope
(c) dark-field microscope
(d) freeze-etch
(e) confocal fluorescence microscope

19. To better view the relationship between tubulin and keratin filaments in NHEKs, you would use _____.

(a) pulse-chase experiment
(b) double-label fluorescent immunocytochemistry
(c) two-dimensional gel electrophoresis
(d) densitometer
(e) affinity chromatography

20. NHEKs actively secrete laminin, a protein that is a component of the basement membrane. To determine if collagen turns on the laminin genes in NHEKs, you would use _____.

(a) native gel electrophoresis
(b) Northern blots
(c) Southern blots
(d) affinity chromatography
(e) gel filtration

21. Laminin is thought to be secreted in tiny (100-nm) vesicles within the cells. What would be the best way to visualize this protein inside an NHEK? _____

(a) ELISA
(b) bright-field microscopy
(c) Nomarski optics
(d) scanning electron microscopy
(e) ultrastructural immunocytochemistry

22. Laminin is a trimeric protein. In other words, it consists of three parts. How might these three parts be best visualized and distinguished one from another? _____
(a) Ion-exchange chromatography
(b) SDS gel electrophoresis
(c) Affinity chromatography
(d) Native gel electrophoresis
(e) Scanning tunneling microscopy

23. During the purification steps of laminin you find that there are some nuclear histones (molecular weight of 14,000–17,000 daltons) contaminating the laminin, whose molecular weight is in excess of 850,000 daltons. Which of the following techniques would be best for further purifying laminin? _____
(a) DEAE ion-exchange chromatography
(b) affinity chromatography
(c) gel filtration
(d) CM ion-exchange chromatography
(e) isoelectric focusing

24. You are interested in examining laminin synthesis in NHEKs. In order to do so you would choose which of the following probes and techniques: _____.
(a) ^{35}S-methionine and gel electrophoresis autoradiography
(b) $^{32}PO_4$ and gel electrophoresis
(c) Fluo3-AM and fluorescence microscopy
(d) micropipets and iontophoresis
(e) rhodamine and ELISA

25. Once you have purified laminin, you then want to visualize it in its native form (i.e., not altered or prepared in any manner). What is the best technique for doing this?_____
 (a) scanning electron microscopy
 (b) freeze-fracture
 (c) freeze-etch
 (d) scanning tunneling microscopy
 (e) light microscopy

26. Explain the importance of paraffin embedding for light microscopy.

27. A dark-field microscope is better for viewing bacteria than a phase microscope with a similar theoretical limit of resolution. Why is this the case?

28. What unique quality of Nomarski (differential interference contrast) optics makes it an attractive microscope system for cell culture biologists who desire to penetrate cells with micropipets?

29. The confocal microscope is quite different from a bright-field or phase microscope. List these differences and explain why the confocal microscope has added to our understanding of cell structure?

30. Fluorescence techniques are increasing in scope in cell biology. Explain why this is the case.

31. Explain how photobleaching of **fluorescent** probes can be used to study plasma membrane fluidity.

32. Explain the relationships between the following:
   (a) antigen; antibody
   (b) rhodamine; fluorescein
   (c) direct; indirect
   (d) monoclonal; polyclonal

33. Define or describe the following:
   (a) pulse-chase
   (b) double-label
   (c) protein A immunoprecipitation
   (d) western blots
   (e) affinity chromatography

34. Compare and contrast the following procedures with emphasis on the advantages and disadvantages of each:
   (a) polyclonal versus monoclonal antibodies
   (b) direct versus indirect technique of fluorescent immunocytochemistry
   (c) transmission electron microscopy versus scanning electron microscopy

35. List the application or technique that would be the most appropriate match for the following problems. In some cases more than one answer is possible. Justify your choice or choices.
   (a) determining which proteins are selectively synthesized in cells in response to **epidermal growth factor**
   (b) determining how much of a particular protein is present in cells
   (c) purifying high-affinity antibodies from a polyclonal mixture
   (d) removing and characterizing a particular protein in solution for which you have an antibody

(e) determining which cells within a heterogeneous culture contain the gene for a particular virus
(f) determining the nucleotide sequence of the **HIV** genome
(g) determining the intramitochondrial location of enzyme X
(h) determining the relative molecular weights of a variety of proteins
(i) determining the intracellular concentration of calcium in epithelial cells in culture
(j) determining where RNA synthesis occurs in the cell
(k) determining if two cells are coupled to each other with an electrical junction
(l) determining if activation of caged molecule X causes an increase in intracellular calcium
(m) determining where within some neural tissue a neuron is located
(n) introducing a gene into a large number of cells to determine its function

36. You purify a protein of interest using native gel electrophoresis and find that its relative molecular weight is 200,000 daltons (200 kDa). You are surprised, however, to discover that when this same protein is analyzed using one-dimensional SDS gel electrophoresis, it now appears as a single band at 100,000 daltons. Not knowing which experiment to believe, you separate the protein using gel filtration and find that it elutes with a relative molecular weight of 200,000 daltons.
   (a) Assuming that the SDS gel electrophoresis data is correct, how might you explain this apparent inconsistency?
   (b) When you treat the 200-kDa protein in question with glutaraldehyde, it now migrates on an SDS gel as a 200,000-dalton species. How might you explain these data?
   (c) You make a monoclonal antibody that can recognize the 200,000-dalton protein. When you attempt a

Western blot using the native gel, you are able to detect the band at 200,000 daltons as expected. When you attempt a Western blot using the SDS gel separation of the aldehyde-fixed protein, no band appears on the Western blot. How might this be explained?

(d) You are now interested in locating the 200 kDa protein in whole cells. You add the antibody to whole cells, followed by a secondary antibody with fluorescein attached to it. You find, however, that there is no signal coming from the cell, suggesting that the antibody didn't bind to the protein. What would you suggest as an explanation?

(e) You now want to determine where this protein is in the cell. The obvious choice for this experiment is to use fluorescent immunocytochemistry or transmission electron microscopy and immunocytochemistry. Neither microscope is working, however. What other choice might you make to answer this question?

(f) How could you use ^{35}S-methionine, protein A immunoprecipitation, SDS gel electrophoresis, and autoradiography to determine the half-life of the protein in the cell?

37. You are doing fluorescent immunocytochemistry using an antibody known to associate with one of the electron transport intermediates in the mitochondria. To your surprise, you find a significant amount of binding of this antibody in the cytosol as well as in the mitochondria. You are convinced that there are two possibilities that could account for this labeling pattern. Either there is a soluble electron transport-like intermediate in the cytosol as well as in the mitochondria, or the cytosolic binding is due to non-specific binding of the antibody. Design an experiment to verify or refute that the cytosolic staining is due to specific binding of the antibody to this electron transport-like intermediate.

Self-Testing Questions

38. How can you distinguish a **cell strain** from a **cell line**?

39. Molecular biology would not be possible without the development of certain enzymes and processes. One of the most important is nucleic acid **hybridization**. List the techniques that have hybridization as a key part of the process.

40. Why is pulse-field electrophoresis sometimes necessary for separating DNA molecules, while it is not necessary for separating protein molecules?

41. What are the similarities and differences between Western, Northern, and Southern blots?

42. Describe the basis of subtractive hybridization. What is an important characteristic of this technique?

43. Explain how hybridization can be used to compare the evolution of species.

44. Why is the heat-resistant nature of *T. aquaticus* so important to the polymerase chain reaction?

ATP

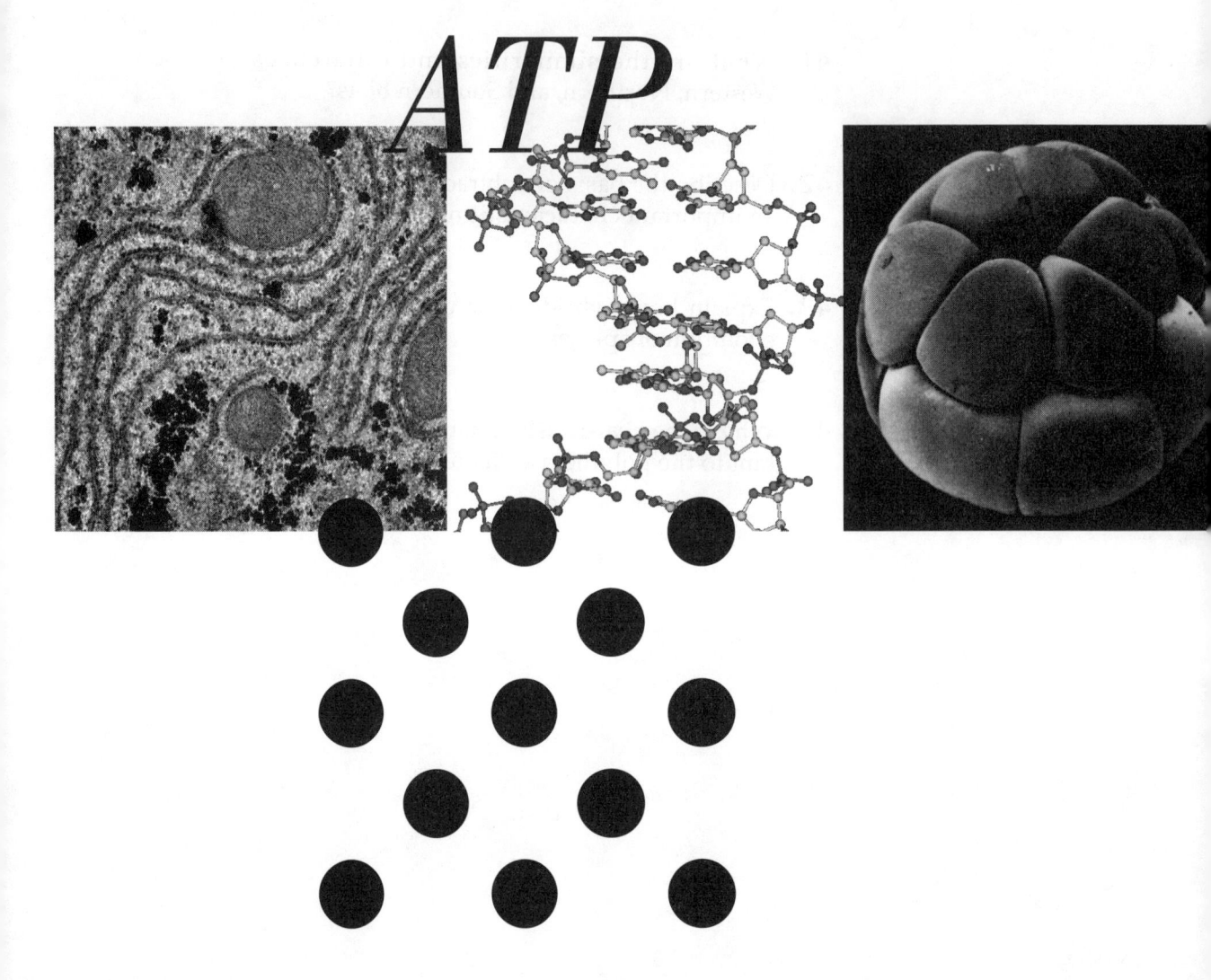

topic 5
Cell Membranes

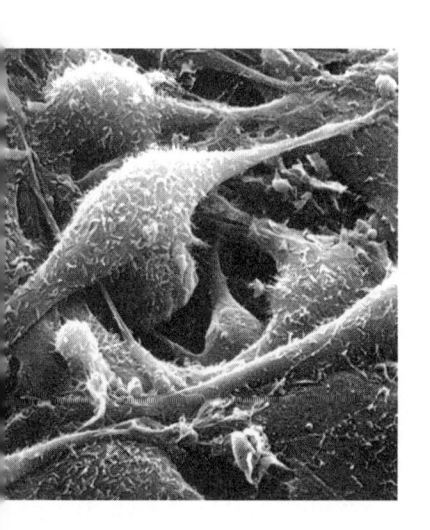

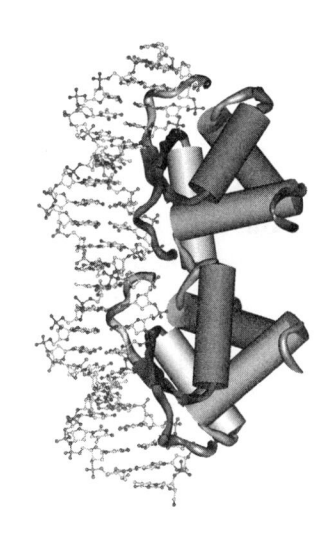

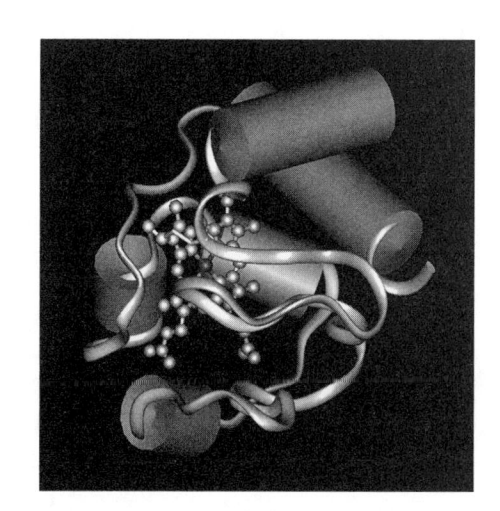

Summary

Cell membranes are the outer limiting barriers of both cells and cellular organelles. Our understanding of the structure and function of membranes has emerged over the last century and has been directly linked to developments in new technology. Even at the beginning of the 1900s scientists knew that cell membranes contained some type of molecule that was **lipophilic** and, as such, would permit lipids, steroids, and other similar molecules to pass through unhindered. Large and charged molecules such as proteins, however, were kept on one side or the other. It wasn't until the pioneering experiments of Langmuir and of Gorter and Grendle in the 1920s that we first realized that cell membranes consist of a phospholipid bilayer. Others such as Danielli and Davson were able to show that proteins were also part of the membrane. This observation was verified by many, such as Bernstein, who removed proteins from cells and found that the electrical resistance increased. But it wasn't until the 1960s and 1970s era that our current **"fluid-mosaic"** model was forged.

Two critical developments during this time resulted in a reappraisal of the original concept of cell membranes, which envisioned a static, protein-phospholipid-phospholipid-protein sandwich model. First, fluorescent probes were being used to tag plasma membrane proteins on living cells. The most notable series of experiments, called **capping and patching**, showed that **antibodies** could bind to membrane proteins and aggregate within the plane of the membrane. It was deduced that the only manner in which this could occur would be if the plasma membrane proteins were mobile—a notion contrary to the then-current membrane theory. Other experiments using lasers (i.e., **fluorescence recovery after photobleaching**) corroborated the idea of mobile proteins. Yet neither of these techniques was able to establish if membrane proteins were merely on the surface of the phospholipid bilayer, consistent with the old theory, or if they penetrated the cell membrane. An electron microscopy procedure was able to unravel this mystery.

Steere, Branton, and others developed the freeze-fracture technique for transmission electron microscopy. This convincingly showed that many cell membrane proteins (e.g., integral, or **transmembrane, proteins**) passed through the membrane. Detergent experiments also showed that some proteins were more localized on the surface and thus easily washed off with soap (peripheral membrane proteins). Model systems such as the red blood cell "**ghost**" were useful in showing that the cell membrane had an asymmetry, with more carbohydrate groups localized on the outside of the membrane than in.

With the advent of transmission electron microscopy in the 1940s, scientists were able to show that plasma membranes had special structures in common with plasma membranes of adjacent cells. For instance, **desmosomes** linked one cell with the next so that they could not be easily separated; **tight junctions** kept fluids in one compartment, and **gap junctions** could pass small molecules such as ions between cells and were responsible for electrical signaling in many cells.

Other work showed that membranes could pass selected molecules through cell membranes but not through **liposomes**—synthetic phospholipid bilayers made exclusively of phospholipids. This was possible through **active transport** pumps such as the sodium-potassium ATPase, as well as passive carriers such as the sodium-glucose pump. Our understanding of plasma membranes also benefited from **ionophores**, molecules that often interpose themselves in membranes. They were used to pass ions such as calcium or sodium down their gradients.

Self-Testing Questions

Match the term on the left with its corresponding partner on the right.

1. Steroid
2. Ionomycin
3. Galactolipid
4. Glycophorin
5. Palmitic
6. Phosphatidylinositol
7. Glycerol

A. Phospholipid
B. **Glycolipid**
C. Fatty acid
D. Integral membrane protein
E. Molecule that can regulate membrane fluidity
F. One of the key components (backbone) of phospholipids
G. An ionophore

Fill in the blanks of the following sentences to make accurate statements.

8. Phospholipids are _____ molecules because they are hydrophobic at one end and hydrophilic at the other end.

9. The presence of phosphate groups is the reason the outside of membranes has an overall _____ charge.

10. Integral membrane proteins can be dislodged only with _____, while peripheral membrane proteins can often be released from the membrane using _____ solutions.

Self-Testing Questions

11. Unsaturated fatty acids have a kink in the molecule owing to the presence of a _____ double bond.

12. _____ is a molecule that can interpose itself between phospholipids and regulate the fluidity of the plasma membrane.

13. When the phospholipid bilayer undergoes a change from a gel-like lipid bilayer to a fluid-like bilayer, it is often referred to as a _____ transition.

14. The opposite of hydrophobic is _____.

15. Bacteriorhodopsin uses _____ to pump _____.

16. Bacteriorhodopsin is present in _____ patches in the bacterium _____.

17. _____ are glycoplipids that are commonly found surrounding or in nerve cell membranes and may serve to separate electrical charge across the membrane better than in a cell membrane without these glycoplipids.

18. A famous experiment by Frye and Edidin fused a mouse cell and a human cell together and tracked the mobility of antibody-labeled integral membrane proteins. This was one of several experiments that led to the proposal of the _____ model of cell membrane structure.

19. Ankyrin in the red blood cell membrane connects _____ and _____.

20. The portion of kidney cells or intestinal epithelial cells that contains microvilli and faces the lumen is called the _____ compartment, whereas the portion of these cells that associates with the basal lamina is often called the _____ compartment.

21. _____ is a membrane-impermeable dye that can be used to unequivocally identify tight junctions.

22. The inside-to-outside membrane potential of most cells (i.e., –70 mV for most nerve cells) can be calculated using the _____ equation.

23. The glucose transporter in red blood cells can distinguish _____ glucose from its stereoisomer _____ glucose and will preferentially transport the _____ isomer.

24. _____ is a drug that can inhibit the function of the sodium-potassium ATPase.

25. The sodium-potassium ATPase is an example of an _____ carrier protein.

26. _____ is the technique that allows researchers to use the transmission electron microscope to view the interior of plasma membranes.

27. The concentration of calcium is _____ outside than inside the cell.

Select the letter that best completes the following statements.

28. A galactolipid is distinguished from other membrane lipids owing to the presence of _____.
   (a) an amino acid
   (b) a nucleic acid
   (c) a carbohydrate
   (d) a sulphate group
   (e) all the above

29. Black membranes are often used to _____.
   (a) test the diffusion characteristics of synthetic phospholipid layers
   (b) determine the tensile strength of cell membranes
   (c) break open cells using detergents
   (d) deliver molecules to the interior of the cells using liposomes
   (e) act as a replica and template for antibodies in the Western technique

30. Glycophorin _____.
   (a) is a **glycoprotein**
   (b) is found in the red blood cell membrane
   (c) is a transmembrane protein
   (d) has hydrophobic amino acids that help anchor the protein in the membrane
   (e) all the above

31. Sodium dodecylsulfate _____.
   (a) is amphipathic
   (b) can form **micelles** in solution
   (c) has a critical micelle concentration above which it can dissolve cell membranes
   (d) is often used to remove integral membrane proteins from cell membranes
   (e) all the above

Topic 5 *Cell Membranes*

32. The experimental protocol that uses fluorescent labels, lasers, and photobleaching to determine the mobility of phospholipids or membrane proteins is referred to as_____.
(a) freeze-fracture
(b) freeze-etch
(c) FRAP
(d) FACS
(e) capping and patching

33. The use of red blood cell ghosts has been a critical procedure in the investigation of cell function. Which of the following has been most often studied using this technique?_____
(a) the relationship between the trans Golgi network and the Golgi
(b) the topology of integral membrane proteins
(c) the variety of mitochondria
(d) the translocation of protein at the level of the endoplasmic reticulum
(e) all the above

34. Fluorescent dyes can be microinjected into cells and often will pass from the interior of one cell to the next. If this is the case and there has been no rupture of the plasma membrane, this dye most probably would have identified _____.
(a) a desmosome
(b) a neurochemical synapse
(c) a tight junction
(d) a gap junction
(e) none of the above

35. Skin cells are very tightly adhered together and as such are replete with _____.
(a) desmosomes
(b) neurochemical synapses
(c) tight junctions
(d) gap junctions
(e) all the above

36. Which of the following is the most permeable (within the group) across a synthetic phospholipid bilayer?_____
(a) calcium ions
(b) magnesium ioins
(c) oxygen
(d) glucose
(e) DNA

37. The sodium-glucose pump is which type of carrier protein?_____
(a) **uniport**
(b) **symport**
(c) antiport
(d) pluriport
(e) none of the above

38. Carrier-mediated transport and diffusion can be distinguished from each other because the former, but not the latter, _____.
(a) is always regulated by ATP
(b) reaches an upper transport rate that cannot be exceeded with the addition of extra cargo (ligand) molecules
(c) uses phospholipids as the primary carriers
(d) occurs only above 15°C
(e) is the only mechanism by which carbon dioxide can enter cells

Topic 5 *Cell Membranes*

39. Most cells, such as neurons, have an inside-to-outside negative charge. This can be calculated by the Nernst equation and is primarily due to the permeability of ____.
   (a) sodium
   (b) calcium
   (c) chloride
   (d) potassium
   (e) manganese

40. The sodium-potassium pump ____.
   (a) pumps sodium out of the cell
   (b) pumps potassium into the cell
   (c) uses ATP
   (d) is an integral membrane protein
   (e) all the above

41. There are a number of agents that can covalently modify and radioactively tag specific, integral cell membrane proteins. They all have the virtue of being impermeable to most cell membranes. Explain how these agents might be used as vectorial labels to determine the topography of an erythrocyte plasma membrane protein.

42. In preparing a population of inside-out and rightside-out **erythrocyte ghosts**, you are concerned about inside-out ghosts contaminating the rightside-out ghost population. How might you use **lectins** to further purify this population?

43. Spectrin is a tetramer that consists of four chains. The α chains have a molecular weight of 240,000 daltons, whereas the β chains have a molecular weight of about 220,000 daltons. Both types can be easily identified on an SDS gel. On a native gel, however, when the tetramer remains intact, the protein appears as the expected 920,000 daltons. When spectrin is incubated with isolated cytoskeletal components from **erythrocytes**, it migrates as a protein with a much larger molecular weight. Explain why this might be the case.

44. Bacteriorhodopsin uses light energy to pump protons from the inside of the cell to the outside. How could you use phenol red to measure its activity?

45. In your attempts to accomplish freeze-fracture of the plasma membrane of cell type X, you notice a rather unusual particle that you suspect might be a large integral membrane protein. Yet you are worried that this particle might be an artifact of your fixation and freezing protocol. How might you assess this possibility?

46. A number of experiments have been used to demonstrate the fluid-mosaic nature of the plasma membrane. Some of the most famous of these experiments are the "patching capping" experiments, in which antibodies bind to integral membrane proteins and eventually form a cap or a patch on the surface of the cell.
(a) In attempting to recreate these experiments, you find that when the antibodies are added, there is no patch formed. Further, *in vitro* analysis shows that the antibody is specific for its target epitope and binding occurs as it should on the living cell, *in vivo*. Give a possible explanation of why this might have occurred as well as a solution to test your theory.
(b) In another series of experiments you add monovalent antibodies to cells and find that patching does not occur. Yet this same antibody in its native, unmodified bivalent form was able to produce patches. Suggest an explanation for this observation.

47. You are investigating the transport of an unusual carbohydrate from outside the cell to the inside. One of the first steps in this investigative process is to determine if the molecule is being transported via active or **facilitated transport**. How might you construct such an experiment?

48. Ionophores are excellent tools to study the effects of changes in intracellular ion concentrations in cells. For instance, ionomycin can cause an increase in intracellular calcium by facilitating the flow of calcium down its gradient from outside the cell (approximately 2 mM) to inside the cell (approximately 100 nM). You discover a new calcium ionophore, and you are not sure if it is an **ion carrier** or a **channel former**. An ion carrier diffuses across the membrane and carries its specific molecular cargo down its electrochemical gradient, whereas a channel former inserts itself into the membrane and forms an ion-specific pore through which ions can flow. How might you use temperature to determine if this new ionophore is an ion carrier or a channel former?

49. Most of the transmembrane proteins have one or two α helices that pass through the plasma membrane. Why is the α helix preferred to the β pleated sheet?

50. Sodium fluorescein is a membrane-impermeable fluorescent dye that has been used to measure tight junction efficacy in a monolayer of kidney cells. Construct an experiment that might use this probe in this manner.

ATP

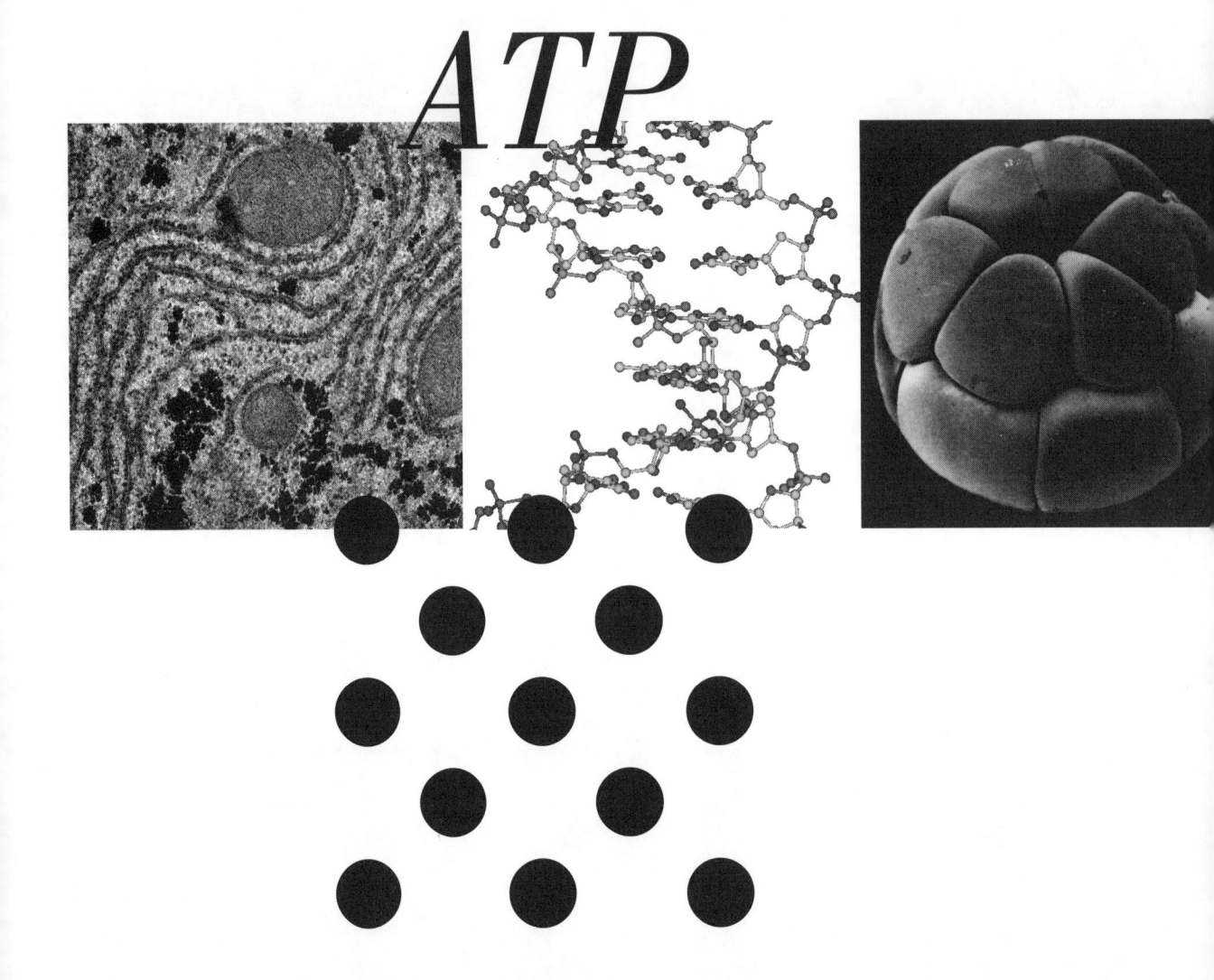

Protein and Vesicular Traffic

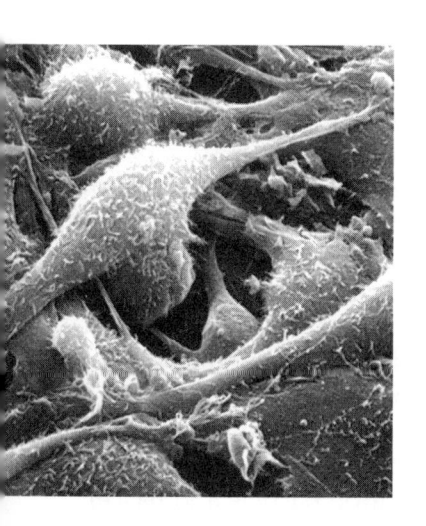

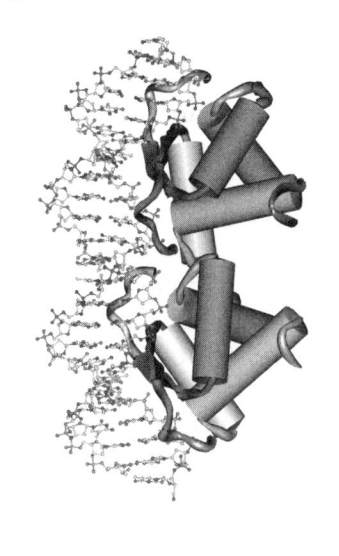

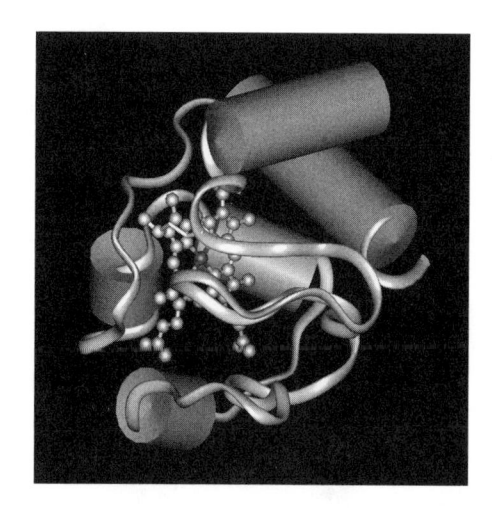

topic 6

Summary

Newly synthesized proteins have multiple destinations in and outside the cell. One of the biggest challenges in the last decade has been to unravel the mysteries of "protein trafficking." Newly synthesized proteins can be (1) free in the cytoplasm, (2) retained by organelles such as the Golgi or the **rough endoplasmic reticulum (RER)**, (3) secreted by cells, (4) shipped to organelles such as the **chloroplast**, mitochondrion, **nucleus**, lysosome, or **peroxisome**, or (5) installed as **integral membrane proteins**. Their respective destinies are determined by where they are synthesized and by which specific "address tags" direct them to future targets.

Proteins are synthesized in two areas in the cell. Many proteins are synthesized on "free ribosomes," ribosomes that are not attached to the endomembrane system. Proteins synthesized here often end up being soluble proteins within the cell cytoplasm. Other proteins synthesized on free ribosomes can be shipped to and incorporated in organelles. But most proteins are synthesized on "bound ribosomes" that are attached to the internal, endomembrane system. Some proteins are synthesized on ribosomes of the outer nuclear membrane, but most are synthesized on the rough endoplasmic reticulum.

Protein synthesis that occurs on the rough endoplasmic reticulum involves multiple steps. Protein synthesis is initiated on free ribosomes and then stops when the **signal recognition particle (SRP)** binds to the complex. The signal recognition particle recognizes the **signal sequence**—the first two dozen amino acids in the nascent protein. This complex of mRNA, ribosome, and signal recognition particle and nascent protein then binds to the rough endoplasmic reticulum—a part of the endomembrane system that originated in evolution from the invagination of the plasma membrane of the protoeukaryote. This complex binds specifically to the rough endoplasmic reticulum through help of the docking protein, also called the **signal recognition particle receptor protein**, as well as the ribosome **receptor**, a different receptor protein. If they are experimentally digested by proteases, the SRP-, mRNA-ribosome complex won't bind.

Proteins are then cotranslationally secreted into the lumen of the rough endoplasmic reticulum if they don't have a **stop-transfer peptide** or **anchor sequence** within the protein. These proteins are then modified in the rough endoplasmic reticulum and the Golgi. They are finally shuttled into the regulatory pathway of the **trans Golgi reticulum** (**trans Golgi network**), stored, and secreted upon command—usually by an extracellular signaling event like hormone binding. Proteins with one or more stop-transfer sequences ultimately become integral membrane proteins.

Protein address tags are also important to protein trafficking. Rough endoplasmic reticulum resident proteins contain a **KDEL** sequence that keeps them bound to a KDEL receptor. Rough endoplasmic reticulum enzymes that have "escaped" to the Golgi are shuttled back to the rough endoplasmic reticulum in transport vesicles. Chloroplast and mitochondrial proteins are synthesized either by nuclear or by chloroplast or mitochondrial genes. Proteins synthesized on nuclear genes often contain two different targeting sequences, one of which targets the outer membrane of the chloroplast or mitochondrion, and a second which targets internal structures within the organelle. Chaperone proteins, **ATP**-dependent enzymes that keep proteins targeted for chloroplasts and mitochondria in an unfolded array, are critical for proper membrane targeting and translocation.

The most-studied and best-known address tag is the mannose 6-phosphate of lysosomal proteins. Enzymes destined for the lysosome are synthesized in the rough endoplasmic reticulum and post-translationally modified with a terminal mannose 6-phosphate. These modified proteins are recognized by mannose 6-phosphate receptors in the trans Golgi network (reticulum) and shipped in a special vesicle to the lysosomes. If the mannose 6-phosphate is not added to this enzyme, as occurs in the human lysosomal storage defect I cell disease, lysosomal enzymes are shipped through the default route to the extracellular space.

In summary, both the site of synthesis and targeting sequences are critically important in protein trafficking.

Self-Testing Questions

Most proteins are translocated through the cell by themselves (nonvesicular transport) after synthesis on free ribosomes, or alternatively, they are packaged in vesicles (vesicular transport). Check the appropriate blank for each of the following newly synthesized proteins to indicate how they are shipped.

Protein Type	Vesicular (secretory pathway)	Nonvesicular (cytosol)
1. Nuclear proteins	_____	_____
2. Lysosomal proteins	_____	_____
3. Peroxisomal proteins	_____	_____
4. Chloroplast proteins	_____	_____
5. Mitochondrial proteins	_____	_____
6. Integral plasma membrane proteins	_____	_____
7. Cytosolic enzymes	_____	_____
8. Cytoskeletal proteins	_____	_____
9. Secreted proteins under regulatory control	_____	_____
10. Secreted proteins under constitutive control	_____	_____

Fill in the blanks of the following sentences to make accurate statements.

11. Proteins with a KDEL sequence are considered _____ proteins within the RER.

Self-Testing Questions

12. _____ is a protein that can catalyze a flip-flop process of lipids.

13. _____ is a polyisoprenoid lipid that is present in the RER and is important for the initial transfer of carbohydrates in many glycosylation processes.

14. The phosphate of the mannose 6-phosphate is added to lysosomal enzymes in the _____ compartment of the Golgi.

15. The pH of the **endosome** budding from the trans Golgi network (reticulum) and heading toward the lysosome becomes progressively more _____ as it approaches the lysosome.

16. Many proteins, such as growth hormones, are released through the calcium-modulated _____ secretory pathway.

17. Glycosylation of proteins can be studied by following a radioisotope's incorporation into nascent proteins during their modification steps in the RER and the Golgi. This type of experiment is called a pulse-_____ experiment.

18. Proteins destined for the nucleus are transported through nuclear _____.

19. The enzyme contained in the RER that clips the signal peptide is called the _____.

20. Some proteins that are firmly associated with membranes but do not contain stretches of hydrophobic transmembrane sequences may be covalently modified through the addition of a _____ anchor.

Select the letter that best completes the following statements.

21. The internalization of LDL, iron, and M6P-addressed lysosomal enzymes is characterized by which of the following processes? ____
(a) phagocytosis
(b) receptor-mediated endocytosis
(c) pinocytosis
(d) autophagy
(e) none of the above

22. Lysosomal enzymes have an address tag that directs them to the lysosome. These enzymes are sorted into their appropriate vesicles in the ____.
(a) RER
(b) mitochondria
(c) cis Golgi
(d) trans Golgi network (reticulum)
(e) coated pit

23. MDCK cells are kidney cells that have been used for classic studies of protein sorting of the VSV glycoprotein and the HA glycoprotein. These cells are excellent candidates for these studies because ____.
(a) they are polarized, with an **apical** and **basal** domain
(b) they grow well in soft agar
(c) they are closely related to 3T3 cells
(d) they multilayer in culture
(e) they are classic cancer cells

24. Which of the following are triskelion molecules associated with vesicle budding from the trans Golgi network heading to the lysosome? ____

(a) KDEL receptors
(b) Bip
(c) histones
(d) clathrin
(e) glycophorin

25. Rough microsomes are structures that were originally ____.

(a) lysosomes
(b) endoplasmic reticulum with ribosomes attached
(c) smooth endoplasmic reticulum
(d) trans Golgi network (reticulum)
(e) none of the above

26. All the proteins targeting the following organelles have signal sequences. All but one have amino-end signal sequences. Select the one that is responsive to carboxyl-end signal sequences. ____

(a) peroxisomes
(b) mitochondria
(c) chloroplasts
(d) RER

27. Nuclear proteins are synthesized in the cytoplasm and then are translocated to the nucleus. They have translocation sequences that are ____.

(a) at the amino end
(b) at the carboxyl end
(c) internally located in the protein
(d) not known

28. The SRP ____.
  (a) binds to the signal sequence
  (b) contains RNA
  (c) contains polypeptides
  (d) can dock with the SRP receptor
  (e) all of the above

29. Bip ____.
  (a) is a chaperone protein
  (b) is found in the rough endoplasmic reticulum
  (c) binds to short hydrophobic sequences of amino acids
  (d) prevents aberrant folding of proteins
  (e) all the above

30. Integral plasma membrane proteins can ____.
  (a) have stop-transfer and **signal-anchor sequences**
  (b) be synthesized in the RER
  (c) be secreted through the constitutive pathway
  (d) have sequences of hydrophobic amino acids or a GPI anchor
  (e) all the above

31. For each drug or agent listed below, indicate the consequence of its action on the protein trafficking pattern of the indicated protein.
  (a) Brefeldin A on secretory proteins
  (b) Nocodazole on receptors that bind to resident rough endoplasmic reticulum (RER) enzymes
  (c) Extracellular concentrations of the calcium chelator EGTA and EGTA-AM on proteins secreted through the regulatory pathway
  (d) Relative differences in effect of protein synthesis inhibitors on secreted proteins of the regulated pathway versus those of the constitutive pathway
  (e) Ammonium chloride or chloroquine on proteins being transported to the lysosome

32. You set up two experimental systems to analyze RER-associated protein synthesis. The components in each system are listed below.

System 1	System 2
ATP, GTP, tRNA	ATP, GTP, tRNA
mRNA coding for collagen	mRNA coding for collagen
Cytosolic enzymes necessary for protein synthesis	Cytosolic enzymes necessary for protein synthesis
Ribosome-free RER (smooth microsomes)	Free ribosomes

You find that both systems are able to produce collagen protein from its mRNA, but that the protein is approximately 2600 daltons larger when synthesized under system 2. Explain why this might be the case.

33. You set up two experimental systems to analyze RER-associated protein synthesis. The components in each are as follows.

System A	System B
ATP, GTP, tRNA	ATP, GTP, tRNA
mRNA coding for collagen	mRNA coding for collagen
Cytosolic enzymes necessary for protein synthesis	Cytosolic enzymes necessary for protein synthesis
Protease-treated microsomes	Microsomes
Ribosomes	Ribosomes

Note that system A has protease-treated rough microsomes, whereas system B does not. In system A, ribosomes were transiently separated from the rough microsome membrane, and the resulting smooth microsomes were treated with protease. Then in both cases these microsomes were added back to the entire solution of enzymes, ribosomes, and mRNA. You find that system A does not generate any collagen protein translocated into the endoplasmic reticulum, whereas system B does generate membrane-bound collagen. How might you explain the difference?

34. You decide to do some initial investigations of a new protein that is being synthesized and shipped to an unknown destination. You have an antibody to this protein so that it can be immunoprecipitated and tracked on a gel system such as a Western blot. You homogenize the cells, spin out the organelles and intact cells and sample the supernatant for the presence of the protein. You find to your delight that the protein is in this supernatant. How might you now determine from this supernatant fraction if your protein is shipped via vesicular transport or nonvesicular transport? **Hint:** Try using proteases in your investigation.

35. You are investigating the ability of a nucleus-encoded mitochondrial protein to target the mitochondria. The intent of your studies is to determine how translocation of this protein occurs. The studies are divided into a number of experimental regimes.
  (a) You find that pretreatment of the mitochondria with proteases abolishes the translocation of this protein. What does this tell you about the mechanism of transport?
  (b) You find that the addition of **oligomycin** does not immediately affect the transport of the protein, but the addition of dinitrophenol (DNP) compromises the ability of the protein to translocate properly. How might this be interpreted?

36. You are tracking a protein that is synthesized on the RER and is transported via the Golgi and trans Golgi network (reticulum) system. You want to determine where the carbohydrate fucose is added. How would you design an experiment to study this?

37. You are tracking a novel protein in a cell and use ultrastructural immunocytochemistry as your sole technique. Surprisingly, you find this protein in the RER, but it is not found anywhere else in the cell. Isolation of the protein from the RER and subsequent amino acid analysis reveals that it has a KDEL sequence. Why is this not surprising?

38. You are confronted with a patient who has a lysosomal storage disease, and you suspect that the problem is with the mannose 6-phosphate (M6P) receptor. You are able to grow some fibroblasts isolated from this patient in culture. How could you use an ^{131}I-labeled lysosomal enzyme to determine *in vitro* if the defect is due to an altered M6P receptor that won't bind normal (wild-type) M6P-addressed lysosomal enzymes?

39. You are following the translocation of a mitochondrial protein and find that you need a cytosol extract in your *in vitro* system to ensure that the protein is properly translocated. Removal of this cytosolic fraction results in no translocation. How might this phenomenon be explained?

ATP

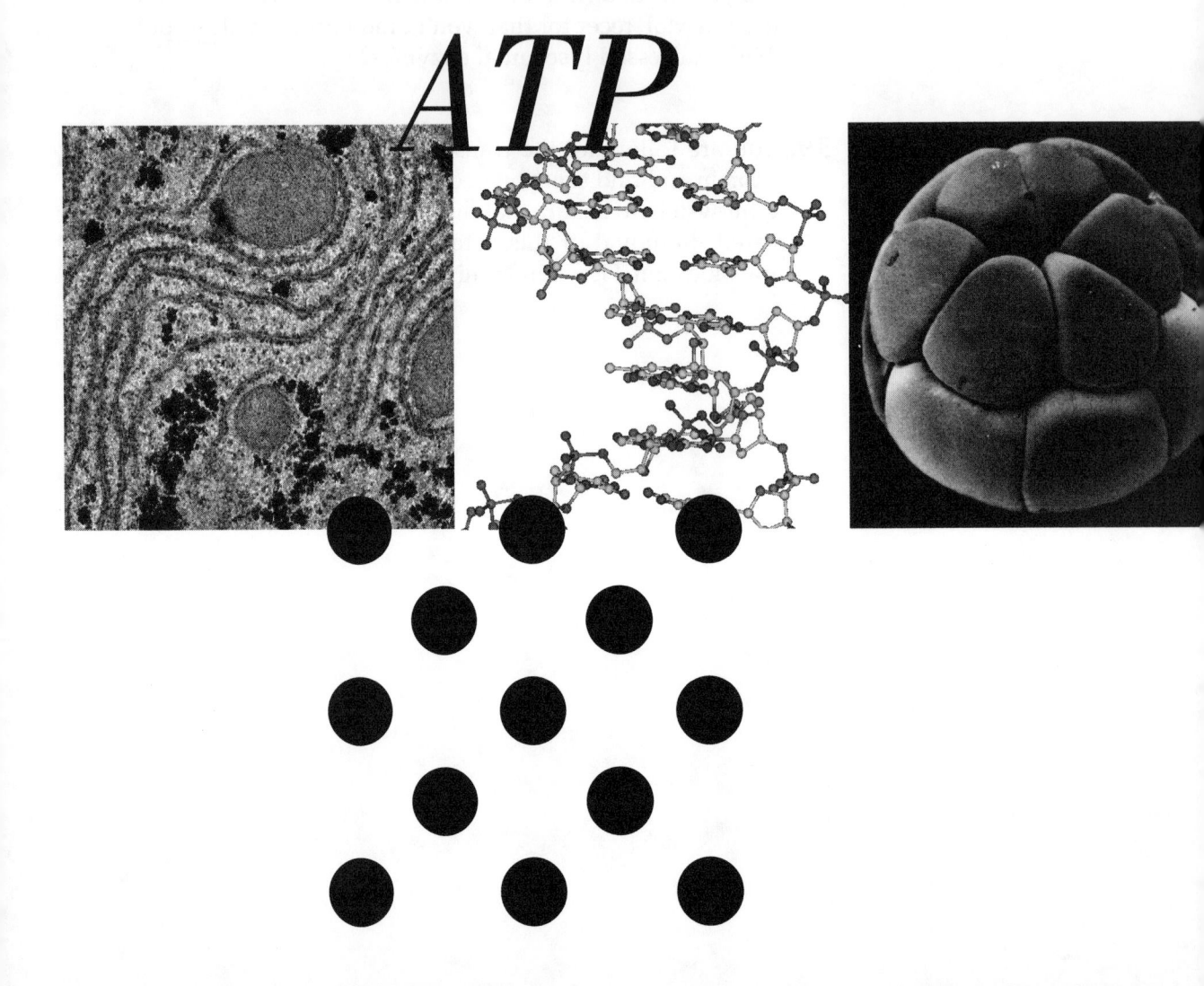

topic 7
Receptors and Second Messengers

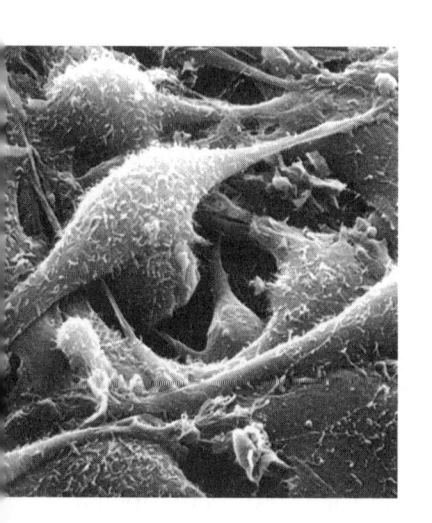

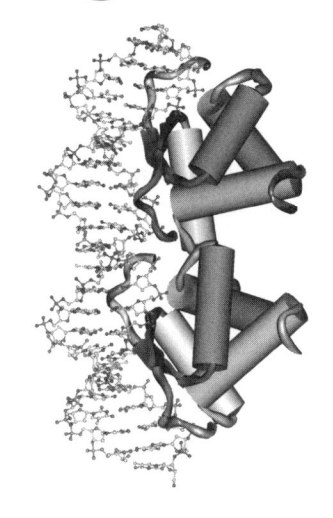

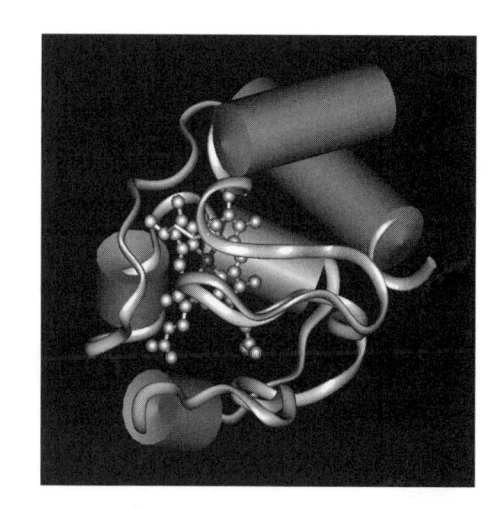

Summary

Essential to multicellular organisms is the capability of cell-to-cell communication over long distances. An intricate system has developed that allows one tissue or organ system to signal other tissues or organs to initiate an activity. This topic deals with the strategy used by cells to accomplish this end.

Cell signaling can occur over short distances (**autocrine** and paracrine signaling), but more typically signaling occurs over long distances—referred to as *endocrine signaling*. Molecules used in this signaling system are referred to as *hormones* and interact with a receptor on the target cell. The mechanism of this action appears to be primarily dictated by the molecular nature of the hormone. For instance, steroids are membrane soluble hormones that can easily pass across the plasma membrane and interact with a soluble cytoplasmic or nuclear receptor. But most hormones aren't membrane-soluble and thus can't signal their intent to the cell directly. They often use some type of signaling cascade.

Hormones such as epidermal growth factor bind to a receptor, causing a dimerization of two receptors. The ligand binding then signals the cytoplasmic tail to become phosphorylated—a process called **autophosphorylation** because the protein kinase that phosphorylates the tyrosine residues is a part of the receptor itself. Next, other proteins such as phosphotyrosine-binding proteins bind to the cytoplasmic tail and phosphorylate a substrate protein that, in turn, effects a change in cell behavior.

Other hormones, such as the neurotransmitters **acetylcholine** and glutamate, bind to a receptor and directly open a channel within that receptor that passes ions down their gradients into the interior of the neuron. The activation of this "ligand-gated channel" can cause a change in the transmembrane potential of the neuron.

Many hormones cause the activation of **second-messenger** systems—relay systems that transfer the message from the membrane-bound hormone receptor to interior destinations within the cell. Several second-messenger systems have been identified, most of which use **G proteins** (i.e., **GTP-GDP-**

binding proteins) as part of the multiple-component cascade. In the adenylate cyclase system, hormones stimulate a receptor, which then binds to a G protein. GTP displaces GDP on the G protein, thus activating the protein. This G protein now activates adenylate cyclase, causing the conversion of **cAMP** from ATP. cAMP, the messenger molecule, then binds to the cAMP-dependent protein kinase in the cytoplasm causing the phosphorylation of proteins. In addition, CREB in the nucleus can be activated by the cAMP system. The protein kinase C system also has a G protein, but the protein kinase that phosphorylates proteins is different from the cAMP-dependent protein kinase. Also, the release of intracellular calcium from the smooth endoplasmic reticulum is linked to activation of protein kinase C.

Finally, other molecules that can't penetrate the membrane can, nonetheless, enter the cell through receptor-mediated **endocytosis**. The ligand binds, clathrin congregates on the interior of the membrane, and a vesicle containing the ligand pinches off from the plasma membrane. This endosome then targets the appropriate organelle [e.g., lysosome if the targeting molecule is low-density lipoprotein (LDL)], and the endosome fuses with the target structure.

Self-Testing Questions

Match the following receptor subtypes on the right with their corresponding descriptions on the left.

1. Has DNA binding region
2. Heterodimer forms from 2 monomers after binding
3. Homodimer forms from 2 monomers after binding
4. Uses clathrin as part of a coated pit
5. Hormone binding can cause a change in transmembrane potential

A. **Receptor tyrosine kinases**
B. Receptor serine-threonine kinases
C. Ion channel receptors
D. Receptor-mediated endocytosis
E. Steroids

Topic 7 Receptors and Second Messengers

Fill in the blanks of the following sentences to make accurate statements.

6. cAMP can be inactivated to 5'-AMP by the enzyme _____, which is inhibited by caffeine.

7. G proteins are regarded as G proteins because they can bind _____ and _____.

8. The affinity of a receptor for a ligand can be calculated using the following equation: _____ = [R][L]/[RL].

9. _____ is a tool used to study the adenylate cyclase system. It is isolated from a bacterium that can cause diarrhea and inhibits the hydrolysis of GTP, increasing cAMP accumulation in cells.

10. cAMP-dependent protein kinases have both _____ units, which inhibit the activity of the enzyme, and _____ units, which phosphorylate substrate proteins.

11. Receptor tyrosine kinases have adapter proteins that link the cyotplasmic domain with another type of signaling molecule. Many of these adapter proteins have an _____ domain that is homologous to a domain in the cytosolic tyrosine kinase encoded by src.

12. Aequorin, Fluo3, and other similar dyes are useful for monitoring the release of _____ from the smooth endoplasmic reticulum during the activation of the **inositol**-phospholipid system.

13. Release of glucagon leads to the increase of blood _____.

Self-Testing Questions

14. Activation of glycogen phosphorylase and inhibition of glycogen synthase lead to the _____ of glycogen.

15. The activation of CREB is due, in part, to the translocation of the cAMP-dependent kinase _____ unit from the cytoplasm to the nucleus.

Select the letter that best completes the following statements.

16. Cells that secrete growth factors affecting target cells far from the secreting cell are regarded as _____.
  (a) autocrine
  (b) endocrine
  (c) paracrine
  (d) all the above
  (e) none of the above

17. Which of the following ligands is regarded as lipophilic? _____
  (a) insulin
  (b) epidermal growth factor
  (c) thyroxine
  (d) LDL
  (e) TGF-β

18. The acetylcholine receptor in neurons passes both sodium and potassium. It is regarded as what type of receptor? _____
  (a) receptor tyrosine kinase
  (b) receptor serine-threonine kinase
  (c) tyrosine-linked receptor
  (d) cAMP activated receptor
  (e) ion-channel receptor

19. The insulin receptor is regarded as what type of receptor? _____

(a) receptor tyrosine kinase
(b) receptor serine-threonine kinase
(c) rough endoplasmic reticulum receptor
(d) tyrosine-linked receptor
(e) ion-channel receptor

20. The number of insulin receptors on a typical cell is approximately _____.

(a) 1
(b) 100
(c) 50,000
(d) 5,000,000
(e) can't be measured

21. The insulin receptor _____.

(a) is a tetramer
(b) has α subunits that bind insulin
(c) can autophosphorylate when stimulated
(d) once activated, can facilitate the internalization of glucose into cells
(e) all the above

22. What occurs if the nonhydrolyzable analog GMP-PNP is presented to homogenates of cells with a fully functional cAMP system? _____

(a) G proteins dimerize
(b) cAMP usually increases
(c) adenylate cyclase is cross-linked with an affinity ligand
(d) the receptors endocytose
(e) the receptors dimerize

23. Isoproterenol is better at binding to the epinephrine receptor than epinephrine is. What does this imply about their K_D's? _____

(a) isoproteronol's K_D is greater than epinephrine's K_D
(b) epinephrine's K_D is greater than isoproternol's K_D
(c) the two K_D's are equivalent
(d) no predictions about K_D's can be made

24. Which of the following G protein subunits in the adenylate cyclase system binds GTP and associates directly with adenylate cyclase? _____

(a) Gs_α
(b) G_γ
(c) G_δ
(d) G_β
(e) none of the above

25. To which of the following molecules is **Ras** most related? _____

(a) clathrin
(b) glycophorin
(c) adenylate cyclase
(d) Gs_α of the adenylate cyclase system
(e) **diacylglycerol (DAG)**

26. The microinjection of which of the following is most likely to block the effects of Ras? _____

(a) anti-Ras antibodies
(b) ATP
(c) AMP
(d) constitutively active RAS protein
(e) calmodulin

27. The sevenless mutant is an actively studied system in _____.
   (a) C. elegans
   (b) R. pipiens
   (c) Drosophila
   (d) X. laevis
   (e) none of the above

28. Which of the following is not part of the MAP kinase cascade? _____
   (a) MEK
   (b) diacylglycerol
   (c) MAP kinase
   (d) Ras
   (e) Raf

29. Calmodulin _____.
   (a) is found soluble in the cytoplasm
   (b) is a protein
   (c) can activate protein kinases
   (d) binds 4 calcium ions
   (e) all the above

30. You are investigating a new receptor for which you have two different analog ligands, A and B. You test them separately in preparations of identical cells and plot the following hormone binding curve.

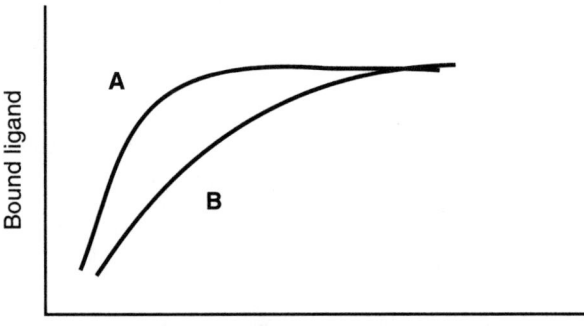

Surprisingly, when you try to displace analog A with analog B you are unable to do so.
  (a) Which analog has the lower K_D (dissociation constant)? How could the actual values for both analogs be calculated?
  (b) Why can't A effectively displace B?

31. One cell line has an epinephrine receptor that increases cAMP accumulation once stimulated. This cell line is then fused with another cell line that has a dopamine receptor linked to adenylate cyclase. Stimulation of the epinephrine receptor in the resulting heterokaryon now results in only a modest accumulation of cAMP. How might this be explained?

32. Affinity ligands are used to identify both receptors and second-messenger components. One such ligand is 8-azido-cAMP^{32}, which can covalently tag the cAMP-dependent protein kinase. This affinity analog acts like cAMP and causes the regulatory catalytic units to dissociate. It forms a covalent bond with the cAMP-dependent protein kinase when irradiated with ultraviolet light. How might this be used to identify the molecular weights of the regulatory subunits of the cAMP-dependent protein kinases?

33. The nonhydrolyzable GTP analog GMP-PNP can cause an increase in cAMP accumulation in extracts of most cells. The addition of cholera toxin can cause an equally large burst in cAMP synthesis. Yet, when both are added together, they are rarely additive in their effects. Why is this the case?

34. You are investigating a novel liver cell line that has an insulin receptor which binds insulin with normal (wild-type) affinity of $K_D = 20$ nM. Yet you suspect that this cell line is unable to transduce the insulin signal.
(a) How could you verify that this cell line is unable to transduce the insulin signal properly into an insulin-dependent change in cell behavior?
(b) How could you determine if the mutant insulin receptor can undergo autophosphorylation of the tyrosine residues in its cytoplasmic domain?

35. Two mutant cell lines have wild-type Ras, but neither can use it to cause a change in cell behavior. Purification of Ras from both cell lines suggests that this G protein is normal and is capable of binding GTP and GDP in a wild-type manner. When the two mutant cell lines are fused, however, the hybrid RAS system works fine. What do you presume might be the deficiencies in these mutants?

36. The inositol-lipid signaling pathway can be stimulated with a combination of the calcium ionophore A23187 and a tumor promoter, phorbel **ester**. Why is this the case?

37. Activation of the epinephrine receptor can cause the phosphorylation of several different proteins through the activation of the cAMP-dependent protein kinase cascade. You think, however, that epinephrine stimulation is causing the phosphorylation of proteins through mechanisms other than the cAMP-dependent protein kinase system as well. How might you discern the difference between these two types of phosphorylation events?

38. A new cell line that you have cloned is unable to respond to stimulation of the inositol-lipid signaling system, whereas its parent cell did. You suspect that one of the problems might be due to the inability of the smooth endoplasmic reticulum to release intracellular calcium in the mutant. How might you investigate this notion?

39. Familial hypercholesterolemia (FH) is characterized by high titers of circulating LDL in the blood. FH can be caused by too few receptors on the cell membrane. You suspect a cell strain isolated from a patient with FH is less able to synthesize new LDL receptors, yet these cells contain the same number of LDL receptors in the cell membrane as do wild-type cells. How might you test this idea?

40. You suspect that the receptor tyrosine kinase EGF is working through the activation of Ras. How might you use anti-RAS antibodies or dominant-negative Ras to confirm your suspicion?

ATP

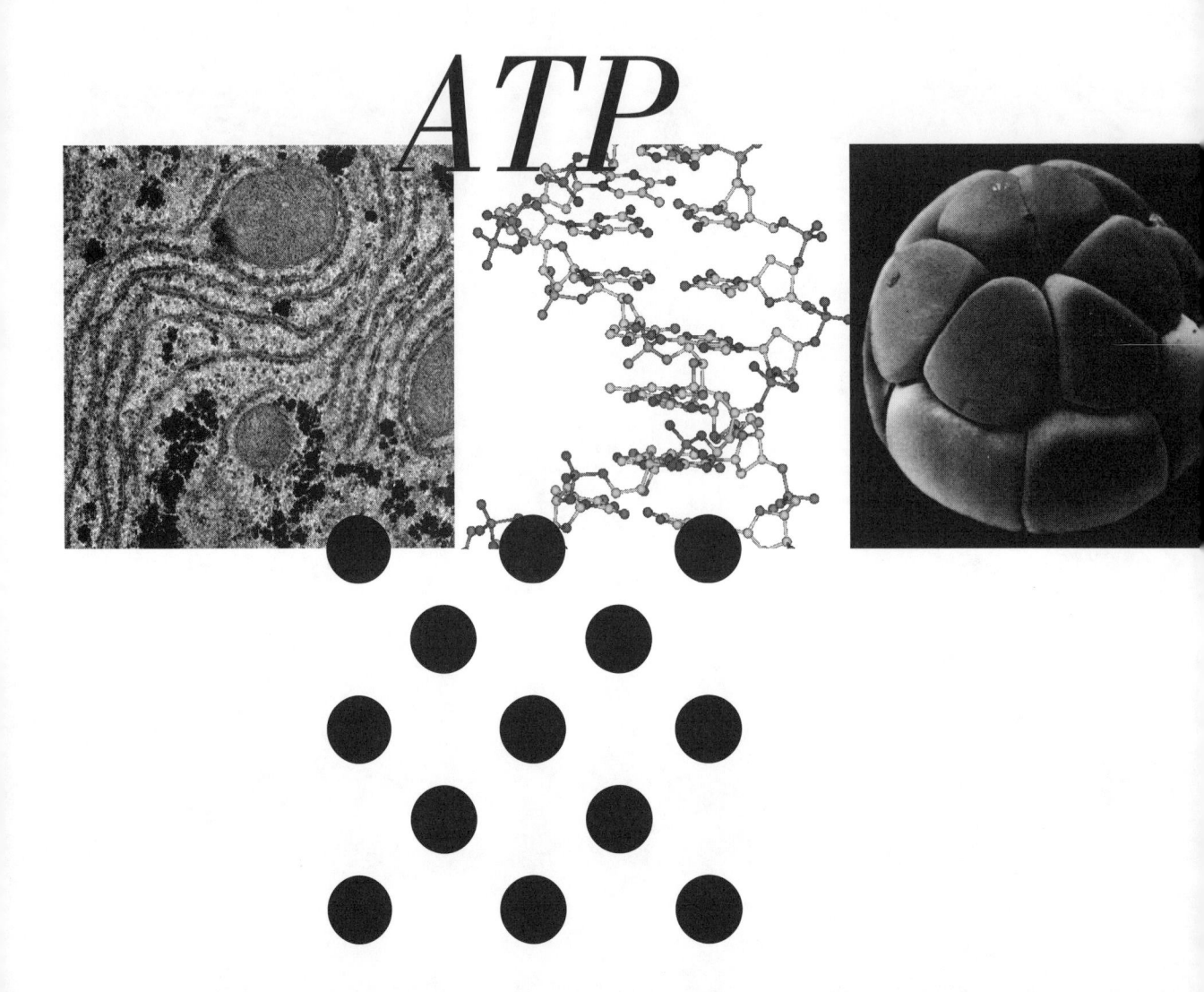

topic 8
Energy, Mitochondria, and Chloroplasts

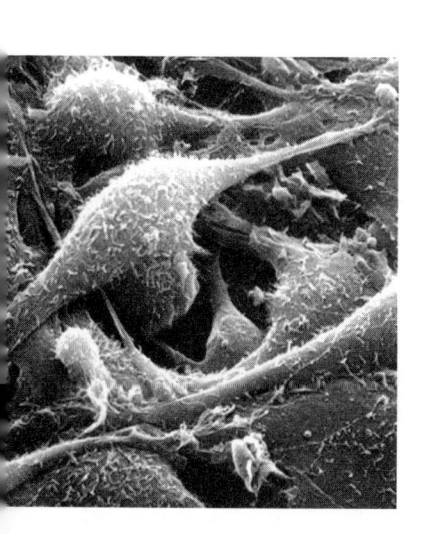

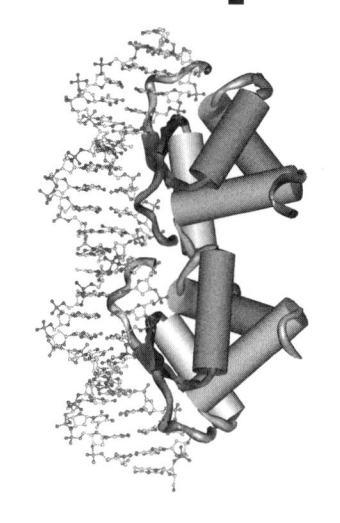

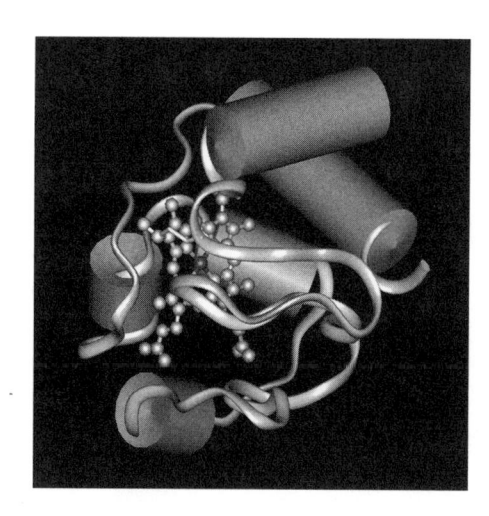

Summary

Cellular energy comes in many different forms. For instance, an asymmetric distribution of ions such as protons or sodium can drive cellular mechanisms directly or indirectly. An example is the sodium-glucose pump. Sodium is present in much higher concentrations on the outside of the cell than on the inside and, via the sodium-glucose pump, glucose can be pumped into the cell against a concentration gradient. The most important type of energy denomination in cells, though, is ATP, *adenosine triphosphate*. ATP is the universal energy currency of the cell and is used for everything from synthesizing proteins to running ion pumps in membranes.

ATP originates from two different places in the eukaryotic cell. Glucose is taken in by the cell and broken down to pyruvate in a process called **glycolysis**. Two ATP molecules are invested in this process to generate glucose 6-phosphate from glucose and fructose 1,6-bisphosphate from fructose 6-phosphate; whereas four ATPs are generated later in the cycle on the path to generating pyruvate. Thus glycolysis, through a process called **substrate level phosphorylation**, is able to generate two molecules of ATP for each molecule of glucose broken down to pyruvate. This is an anaerobic process; that is, oxygen is not necessary for this cycle to proceed. Yet if animal eukaryotic cells were to rely on glycolysis exclusively for their ATP needs, over 90% of the energy-demanding activities of the cell would come to a halt! Specialized organelles, called *mitochondria*, can provide the necessary ATP to meet these additional energy demands of the cell.

Mitochondria are unusual organelles in many ways. First, they have two, not one, delimiting membranes. The two membranes are absolutely necessary for ATP production. Second, they appear to have evolved from an ancestral prokaryote. The evidence for this origin comes from many different fronts. For instance, mitochondria have ribosomes that are structurally more similar to prokaryotes than eukaryotes. ATP is generated in mitochondria through chemiosmosis. Pyruvate or fatty acids enter the mitochondria from glycolysis and are channeled into the critic acid cycle (also called the **Krebs** cycle, or *tricarboxylic acid cycle*). Carbon dioxide, **NADH**, and FADH$_2$ are released as a consequence of the complete breakdown of acetyl CoA. While the NADH and FADH$_2$ are energy-rich molecules, they are not in the correct "currency" for the cell. They are converted to ATP through chemiosmosis.

NADH gives up its electrons to the **electron transport system** in the mitochondria. This shuttle system yields **NAD + H⁺**. The resulting proton is pumped across from the matrix to the intermembrane space. The same process occurs with FADH$_2$. The result is both an electrical potential (−200 mV, with the matrix negative) and a pH differential that can drive ATP synthesis. Protons now move down their gradient from the intermembrane space to the matrix through the ATP synthase, otherwise known as the F_0F_1 *complex*. This results in the phosphorylation of ADP to ATP. Much of our knowledge of the actions of mitochondria comes from using different poisons such as cyanide, an electron transport inhibitor; dinitrophenol, an electron transport uncoupler that compromises the pH differential; valinomycin and gramicidin, antibiotics that compromise the electrical potential; and oligomycin, an agent that inhibits the F$_0$F$_1$ complex.

Chloroplasts also generate ATP but in a different manner and for a different purpose. Chloroplasts are larger than mitochondria, but like mitochondria, probably originated from an ancestral prokaryote. The chloroplast has two outer limiting membranes. Unlike mitochondria, though, it also has closed inner-membrane sacs in the inner stroma. These thylakoid membranes are the sites of **photophosphorylation**. Light interacts with **chlorophyll** molecules in the thylakoid membranes either directly or indirectly through auxiliary pigments. Light causes chlorophyll to release a pair of electrons that subsequently pass down the photosystem II electron transport chain. Next, light causes the electrons to be released from photosystem I, with NADP⁺ + H⁺ being the final electron acceptor, forming NADPH. A proton gradient is created, with the thylakoid lumen being more acidic than the outside stroma. Protons are then able to flow down their gradient through ATP synthase, otherwise known as the CF_0F_1 *complex*. ATP is generated as a result. This ATP and the NADPH produced from this cycle are then used in the stroma's Calvin cycle, sometimes known as *carbon fixation*, to generate sugars from carbon dioxide. The key enzyme of the Calvin cycle is **RUBISCO**, ribulose 1,5-bisphosphate carboxylase, a somewhat sluggish enzyme with a low affinity for carbon dioxide. It also has a wasteful oxygenase activity that is counterproductive to carbon fixation.

Self-Testing Questions

Indicate whether the following statements are True or False.

1. The energy currency of the cell is ATP. Hydrolysis of the terminal high-energy phosphoanhydride bond results in a free-energy change of −7.3 kcal/mol.

2. In order for ATP to be synthesized during glycolysis, the reaction to which the phosphorylation of ADP to ATP is coupled must by endergonic.

3. ATP is generated in glycolysis through **chemiosmotic phosphorylation**.

4. The citric acid cycle is also called the *Krebs cycle*.

5. The citric acid cycle occurs in the intermembrane space of the mitochondria.

6. Cardiolipin in the inner mitochondrial membrane enhances the permeability of that membrane to protons.

7. Bacteria can pump protons and generate ATP through chemiosmotic phosphorylation. This generates an acidic cellular interior, relative to the outside.

Self-Testing Questions

8. The electrical potential across the inner membrane in actively respiring mitochondria is approximately −200 MV.

9. The ability of fructose 2,6-bisphosphate to stimulate phosphofructokinase-1 is referred to as a positive allosteric effect.

10. Some of the energy released during the catabolism of glucose is released as heat.

11. Purple bacteria use two photosystems.

12. PSII, but not PSI, splits water, releasing oxygen, protons, and electrons.

13. P_{680} is associated with PSI.

14. Plastocyanin is a soluble electron carrier in chloroplasts.

15. C_4 plants are often found in cool, moist environments such as mountains.

Fill in the blanks of the following sentences to make accurate statements.

16. Bacteria that can't grow in the presence of oxygen are regarded as _____.

17. Bacteria that can grow either in the presence of oxygen or without oxygen are regarded as _____.

18. Oxygen becomes limiting during high-energy output of mammalian muscles. As a result, pyruvic acid is converted to _____.

19. Bacteria, mitochondria, and chloroplasts all produce ATP through _____.

20. The energy potential generated across membranes in bacteria, mitochondria, and chloroplasts that can generate ATP is regarded as the _____-motive force.

21. _____ is the enzyme that is able to phosphorylate glucose to glucose 6-phosphate.

22. ATP is produced during glycolysis when phosphoenolpyruvate is converted to _____.

23. _____ help to expand the surface area of the inner mitochondrial membrane.

24. _____ is the final electron acceptor in mitochondrial electron transport.

25. Fatty acids are metabolized in the mitochondria by first being converted to _____.

26. The oxidation of NADH to NAD^+ is _____ exergonic than the oxidation of $FADH_2$ to FAD. (Note: choose either "more" or "less.")

27. The pH of the mitochondrial matrix is _____ than the pH of the intermembrane space. (Note: choose either "higher" or "lower.")

Self-Testing Questions

28. The F_0 portion of the F_0F_1 complex in mitochondria contains _____ **subunits**.

29. Isolated F_0 subunits can _____ ATP.

30. The F_0F_1 complex is often referred to as the ATP _____.

31. The ATP/ADP **antiport** protein is located in the _____ mitochondrial membrane.

32. The malate shuttle transports _____ from the cytoplasm, where it is produced in glycolysis, to the mitochondrion, where it can be used to generate ATP through oxidation by the electron transport chain.

33. When one molecule of NADH is oxidized to NAD^+, _____ electron(s) and _____ proton(s) are released. (Note: choose appropriate numbers.)

34. _____ is the only electron carrier that is not a protein-bound prosthetic group.

35. Cytochromes contain a _____ prosthetic group.

36. Fatty acids can be oxidized in the _____ without the production of ATP.

37. Photosynthesis occurs on _____ membranes, whereas carbon fixation occurs in the _____ of chloroplasts.

Topic 8 *Energy, Mitochondria, and Chloroplasts*

38. Chlorophyll is a ringed compound containing the divalent cation _____.

39. Chlorophyll, excited by light, removes electrons from _____.

40. The _____ the wavelength of light, the lower the energy.

41. _____ constant is an important consideration in calculating the energy of a photon.

42. **Photorespiration** is generally regarded as a wasteful process because it liberates _____ and uses _____.

43. _____ is the key enzyme that catalyzes the conversion of phosphoenolpyruvate into oxaloacetate in C_4 plants, thus incorporating carbon dioxide.

Select the letter that best completes the following statements.

44. Glucose _____.
  (a) is transported from the cytoplasm into the mitochondria
  (b) is converted to fatty acids before entering the mitochondria
  (c) is broken down to pyruvate in the cytoplasm
  (d) is a disaccharide
  (e) all the above

45. Which of the following is *not* found in the mitochondria? _____

(a) calcium granules
(b) DNA
(c) ribosomes
(d) electron transport system
(e) none of the above; that is, all are found in the mitochondria

46. Which of the following is characteristic of the outer mitochondrial membrane, but not the inner mitochondrial membrane? _____

(a) high concentration of integral membrane proteins
(b) presence of porin
(c) presence of electron transport system
(d) greater permeability
(e) higher resistance to detergent solublization

47. Glycolysis _____.

(a) is a catabolic process
(b) refers to the breakdown of glucose
(c) is an anaerobic process
(d) yields a net 2 ATP per molecule of glucose
(e) all the above

48. Which of the following enzymes in glycolysis splits fructose 1,6-bisphosphate to dihydroxyacetone phosphate and glyceraldehyde 3-phosphate? _____

(a) aldolase
(b) hexokinase
(c) phosphoglyceromutase
(d) phosphoglycerate kinase
(e) pyruvate kinase

49. NADH is produced by glycolysis through the action of glyceraldehyde 3-phosphate dehydrogenase. This NADH can then be shuttled into which of the following organelles or processes, where it can be converted to ATP? _____

(a) peroxisomes
(b) glycolysis
(c) mitochondria
(d) lysosomes
(e) nucleus

50. Which subunit of the F_1 complex binds ADP and P_i to form ATP? _____

(a) α
(b) β
(c) γ
(d) δ
(e) ϵ

51. Which of the following enzyme(s) is important for substrate-level phosphorylation of ATP from ADP? _____

(a) phosphoglycerate kinase
(b) pyruvate kinase
(c) enolase
(d) hexokinase
(e) both (a) and (b)

52. Which of the following structures appears as a "lollipop-like" figure in negatively stained electron micographs? _____

(a) NADH
(b) $FADH_2$
(c) F_0F_1
(d) cytochrome c oxidase
(e) acetyl CoA

53. Fatty acids can be converted to more ATP than glucose can on a gram per gram basis. Fatty acids are stored in _____.
(a) starch
(b) glycogen
(c) peptidoglycans
(d) triacylglycerols
(e) none of the above

54. Which of the following is *not* an intermediate in the Krebs cycle? _____
(a) succinate
(b) fumarate
(c) isocitrate
(d) oxaloacetate
(e) fructose 6-phosphate

55. The total oxidation of glucose that occurs in both glycolysis and the Krebs cycle results in the release of how many molecules of carbon dioxide? _____
(a) 2
(b) 3
(c) 4
(d) 5
(e) 6

56. Most of the energy produced from glycolysis that can ultimately be converted to ATP is contained in _____.
(a) NADH and FADH$_2$
(b) the electron transport intermediates
(c) water
(d) GTP from the Krebs cycle
(e) carbon dioxide

57. The chemiosmotic hypothesis of **oxidative phosphorylation** was proposed by _____.
(a) Daniel Branton
(b) Rita Levi-Montalcini
(c) Peter Mitchell
(d) Singer and Nicholson
(e) Schleiden and Schwann

58. Bacteriorhodopsin incorporated into synthetic membranes containing F_0F_1 particles is able to generate ATP because bacteriorhodopsin _____.
(a) pumps protons across the membrane
(b) complexes with the F_0F_1 particles
(c) blocks the proton flow through the F_0F_1 particles
(d) can help invert the membranes
(e) all the above

59. The comparison of the action spectrum and the absorption spectrum supports _____.
(a) the presence of the electron transport system in chloroplasts
(b) the hypothesis that the **proton-motive force** drives ATP synthesis in chloroplasts
(c) the idea that chlorophyll, carotenoids and other major pigments are directly related to photosynthesis
(d) the rationale for C_4 plants
(e) mitochondrial oxidative phosphorylation

60. Chlorophylls in the reaction centers do not absorb enough light to fully support photosynthesis. The other contributing factor is _____.
(a) ATP generated from mitochondria
(b) light-harvesting complexes containing chlorophyll b and other auxilliary pigments
(c) the limitation of water
(d) the presence of the CF_0F_1 particle
(e) none of the above

61. RUBISCO _____.
   (a) is the abbreviation for ribulose 1,5-bisphosphate carboxylase
   (b) catalyzes the fixation of carbon dioxide into carbohydrates
   (c) is one of the enzymes of the carbon fixation cycle
   (d) occurs in large amounts in plants
   (e) all the above

62. The final cytosolic product of carbon fixation is _____.
   (a) sucrose
   (b) ribulose 5-phosphate
   (c) 1,3-bisphosphoglycerate
   (d) glucose
   (e) fructose 1,6-bisphosphate

63. One of the earliest experiments on chloroplasts recorded at the turn of the century was the observation of isolated chloroplasts in a sucrose solution. Bacteria that contaminated the sucrose solution appeared to associate preferentially with the isolated chloroplasts rather than range freely within the sucrose medium. Why was sucrose, rather than pure water, used in this experiment? Also, why were the bacteria preferentially associated with the chloroplast membrane?

64. Thylakoid membranes can be isolated from chloroplasts intact and analyzed for activity. These preparations are useful for exploring the nature of the chemiosmotic hypothesis underlying photophosphorylation.
   (a) Why would phenol red be a good choice as a tool to monitor the activity of thylakoid membranes?
   (b) When these isolated thylakoid disks are illuminated by light of two different wavelengths (680 and 700 nm), the change in color of phenol red is greater than when the thylakoid disks are illuminated with the summed amplitude of the single, averaged wavelength, 690 nm. Why is this true?

(c) The intralumenal pH of thylakoid membranes can be adjusted by soaking them in media of different acidities. By adjusting the pH as indicated below, you obtain the following data. Explain the results:

Intralumenal pH	Outside pH	Result
4.0	4.0	No ATP synthesized or hydrolyzed
8.0	8.0	No ATP synthesized or hydrolyzed
8.0	4.0	ATP is hydrolyzed

(d) When radioactively tagged chloride ions are added to these isolated thylakoid membranes, the isotope passes readily into the interior of the thylakoid disk. If a similar experiment is done with closed, inner mitochondrial membranes, chloride does not pass as easily into the matrix of the mitochondria. How might this be explained and what is the implication for the proton-motive force differences between the two systems?

65. How can the multiprotein complexes in the thylakoid membranes best be visualized and distinguished from one another?

66. Since photophosphorylation in chloroplasts can produce ATP that is utilized for cellular metabolic reactions, why do plant cells contain mitochondria?

67. You are analyzing a new putative mitochondrial poison that you suspect might poison the pyruvate transporter. How might you set up an experimental system to test this hypothesis?

68. An apparent paradox in the chemiosmotic hypothesis is the fact that the pH of the intermembrane space of mitochondria is the same as that of the cytosol, even though the electron transport system pumps protons from the matrix to this intermembrane space. How can this paradox be reconciled?

69. The chemiosomotic hypothesis has been tested in a variety of methods. One of the most useful has been the use of inside-out submitochondrial vesicles (particles). These vesicles are prepared by first isolating mitochondria through **centrifugation** techniques and then stripping off the outer membrane. Next, ultrasound is used to produce an experimental vesicle that appears as shown below.

(a) Locate the F_0F_1 particle in this inside-out mitochondrial vesicle.
(b) In which direction are the protons being pumped?
(c) The addition of dinitrophenol, a proton ionophore, does not compromise the activity of the electron transport chain but does decrease oxidative phosphorylation. Why is this the case?
(d) The addition of valinomycin, a potassium ionophore, does not compromise the activity of the electron transport chain but does decrease oxidative phosphorylation. Why is this the case?
(e) The addition of sodium azide, an electron transport inhibitor, decreases oxidative phosphorylation in the inside-out vesicles. How might this be explained?
(f) Oligomycin causes an increase in proton-motive force in these inside-out submitochondrial vesicles, yet oxidative phosphorylation is inhibited. Explain this apparent paradox.
(g) Chloramphenicol is added to the inside-out submitochondrial vesicles, and oxidative phosphorylation does not appear to be affected. Why is this the case?
(h) Treatment of inside-out submitochondrial vesicles with small amounts of biological detergents, such as sodiumdodecyl sulfate, causes a decrease in oxidative phosphorylation. Why might this be the case?

(i) Mechanical disruption of the inside-out submitochondrial vesicles can lead to dissociated F_1 particles that are no longer associated with the F_0. In this case electron transport continues but no ATP is synthesized. Explain why this is true.

(j) When the F_1 particle is removed from the F_0 and the F_0 is left in the membrane of the inside-out particles, the proton-motive force dissipates, but the electron transport system continues to operate. The outside medium becomes acidic. What does this demonstrate about the F_0 particle?

70. Rhodamine 123 is a vital dye that can label living mitochondria. When this fluorescent dye is added to actively respiring cells, mitochondria appear fluorescent. When mitochondrial activity is increased, the mitochondria appear even more fluorescent. Initial studies of this dye, using whole living cells and rhodamine 123, yielded the following data.

Treatment	Drug Action	Effect on Rhodamine 123 Retention
Chloramphenicol	Inhibitor of mitochondrial protein synthesis	No effect
Cycloheximide	Inhibitor of cellular protein synthesis	No effect
Sodium azide	Electron transport inhibitor	Decrease
Dinitrophenol	Proton ionophore	Decrease
Valinomycin	Potassium ionophore	Decrease
Oligomycin	F_0F_1 inhibitor	Increase

What do the data reveal about rhodamine 123 retention?

ATP

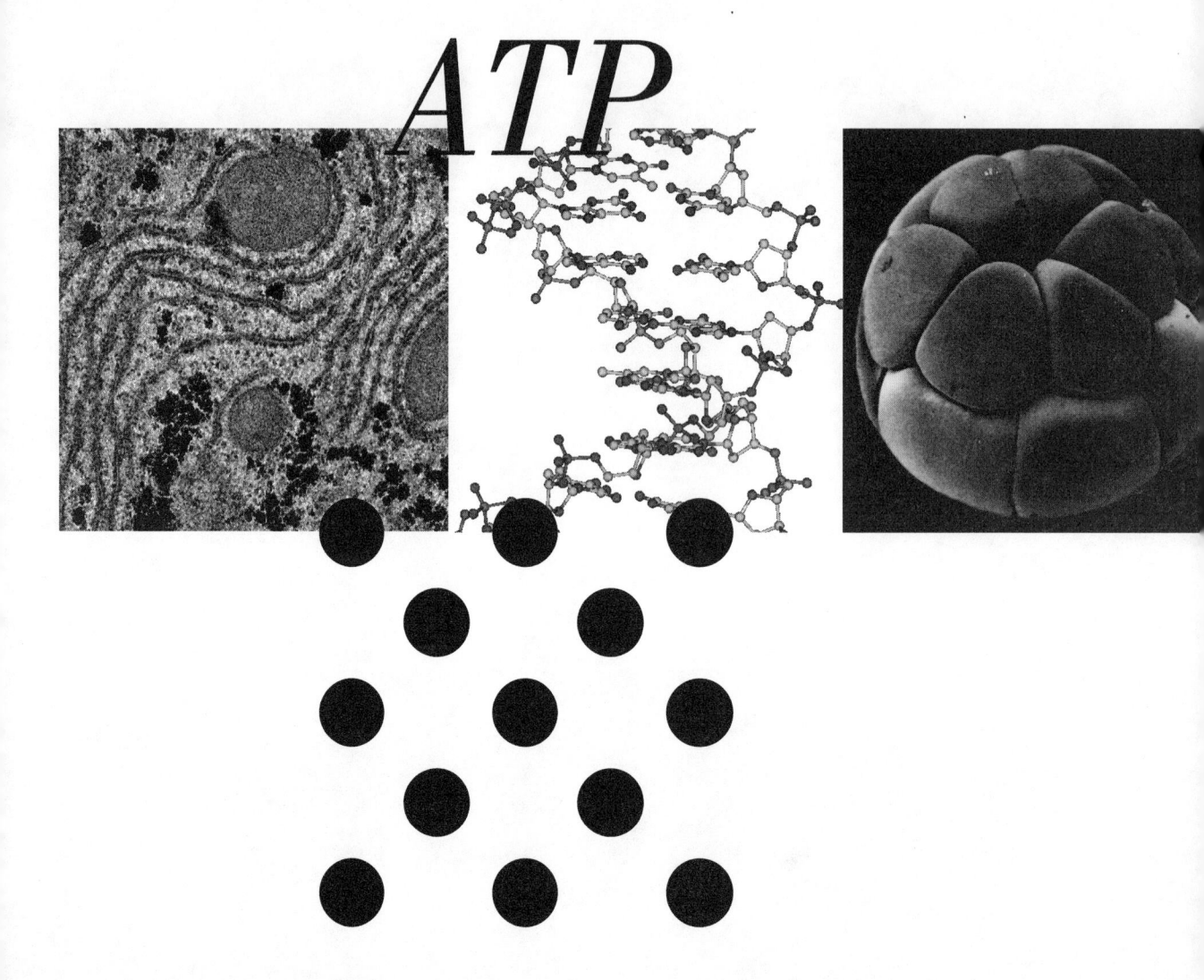

topic 9
Cytoskeleton and Cell Movement

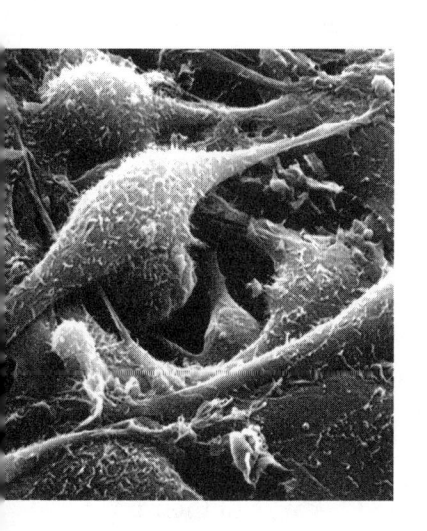

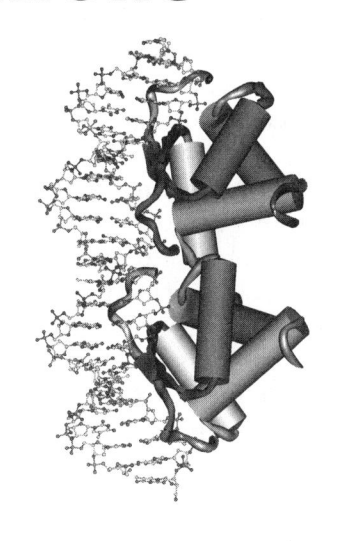

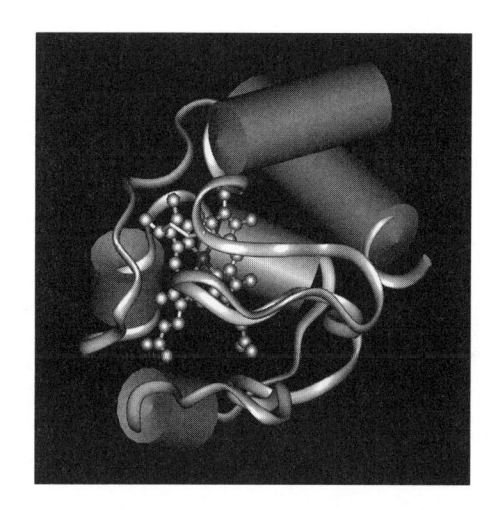

Summary

The cytoskeleton is the internal framework of the cell. It has many different functions including maintenance of structure and movement and transport of molecules. There are three major elements that comprise the cytoskeleton. In order of size from smallest to largest they are actin, intermediate filaments, and **microtubules**.

Actin filaments are found primarily around the periphery of the cell. For decades it has been known that they are the internal structure of microvilli, small extensions of the cell that increase the absorptive area. Similarly, it has been recognized that cell movement is mediated by the assembly and disassembly of actin filaments. Both these facts can be proved through the addition of cytochalasin, an actin destabilizer. If this inhibitor is added to living cells, the microvilli collapse into the cell, and cell movement of ameboid cells is inhibited.

Actin is a highly conserved protein and is present at very high concentrations in eukaryotic cells. Approximately half of the actin is in monomeric form, known as *G (globular) actin*, whereas the other half of the actin is present in the polymeric form, known as *F (filamentous) actin*. F actin, like microtubules, has a polarity, with the ends designated as (+) and (−), respectively. The conversion between G actin and F actin is mediated by ATP. This interchange is a dynamic process that gives rise to microspikes and lamellipodia, the "legs" of a moving cell. There are several proteins in cells that regulate actin assembly. For instance, thymosin binds monomeric G actin in cells and thus can partially regulate the amount of F actin.

Intermediate filaments are so named because of their intermediate length between actin and microtubules. They are extremely strong, compared with other cytoskeletal elements, and have a rope-like structure. In monomeric form they consist of rods that form coiled-coil structures when the protein forms a dimer. Two dimers then form a tetramer—the major structural element of the intermediate filament. Unlike actin filaments and microtubules, intermediate filaments are not polarized. Intermediate filaments are particularly important in situations where tensile strength is critical. They also come in a variety of types. For instance, keratin filaments are found in human skin cells, neurofilaments are found in nerve cells, **lamins** are the key structural elements of the nucleus, and desmin filaments are found in **muscle cells**.

Microtubules are present in most cells and are the internal structural elements for cilia. They are the primary cytoskeletal element of overall cell structure. The addition of colchicine or nocodazole, two microtubular destabilizing agents, causes the cell to collapse upon itself. Microtubules also act as a framework that organizes the position of the Golgi and the rough endoplasmic reticulum. Finally, microtubules act as the guides or struts on which molecules are moved from one place in the cell to the next.

Microtubules consist of tubulin dimers, composed of α-tubulin and β-tubulin. The polymerization of the tubulin dimers to microtubules is governed by both GTP and microtubule-associated proteins (MAPs). This organization creates a polarity. The (–) end of the microtubule is most often located near the center of the cell, whereas the (+) end is usually located nearer the periphery. The addition of the tubulin dimers to either end can be governed by the critical concentration of tubulin within the cell. The greater the concentration of tubulin dimers, the higher the chance that the microtubule will be formed. The (–) end is fundamentally different from the (+) end, since it is more unstable. Thus, in most cells the (–) end is "capped" so that tubulin dimers are not lost from this end, causing a depolymerization of the organelle. Kinesin is a molecular motor that binds to microtubules and can transport molecules to the (+) end of the microtubule, whereas dynein is molecular motor that transports molecules to the (–) end of the microtubule.

Self-Testing Questions

Match the protein on the left with its function on the right.

1. Tropomyosin
2. Gelsolin
3. Spectrin
4. **Myosin** II
5. Myosin I
6. α-Actinin
7. Filamin
8. Dystrophin
9. Profilin
10. Titin

A. Coded by gene with 2 million bases and associated with skeletal muscle

B. Attaches actin filaments (indirectly) to plasma membranes

C. Large, spring-like molecule in skeletal muscle

D. Mediates vesicle movement on actin filaments

E. Splits actin filaments into smaller pieces

F. Makes actin filaments stronger

G. Cross-links actin filaments

H. Binds to actin filaments and makes them slide together

I. Binds actin monomers in a 1:1 ratio making it unable to bind to filament

J. Bundles actin filaments together

Fill in the blanks of the following sentences to make accurate statements.

11. The combination of actin, intermediate filaments, and microtubules forms the _____ of cells.

Self-Testing Questions

12. _____ is the most abundant intracellular protein in eukaryotic cells.

13. _____ is often used to "decorate" actin filaments so that polarity of the actin filaments can be revealed.

14. The decoration referred to in question 13 is best seen using _____ microscopy.

15. _____ is a very large protein found missing in people with Duchenne muscular dystropy. It anchors actin filaments to the **extracellular matrix**.

16. The plus end of actin filaments grows _____ than the minus end.

17. Depolymerizaton or polymerization of actin filaments can be regulated in the test tube by diluting or concentrating (respectively) the concentration of G-actin in solution. In this case one is determining the _____ of actin necessary to achieve polymerization.

18. Kinesin is to microtubules as _____ is to actin.

19. The individual force of a myosin molecule can be measured using an _____ trap.

20. Smooth muscle differs from skeletal muscle in lacking the _____ characteristic of skeletal muscles.

21. A unit of muscle from one Z disk to the next is referred to as a _____.

22. The **sarcomere** contains thick filaments, which are _____ filaments.

23. Nebulin is a protein that is very closely associated with _____ filaments in skeletal muscle and may serve to keep them in register with one another.

24. Actin filaments that terminate in an adhesion plaque are indirectly associated with the extracellular matrix protein _____.

25. GTP binds to the _____ tubulin unit and hydrolyzes during the process of microtubular elongation.

26. Microtubules usually consist of 13 _____.

27. Hooked-shaped arms of _____ can assemble under high salt conditions and thus demonstrate the polarity of microtubules.

28. The _____ is sometimes known as a *centrosome* and serves as a nucleating or stabilizing center for microtubules.

29. A _____ is similar to a centrosome, but contains only one centriole.

30. Addition of colchicine or nocodazole can depolymerize _____.

Self-Testing Questions

Select the letter that best completes the following statements.

31. The cytoskeleton consists of three different types of filaments. Ordered from smallest to largest they are _____.

(a) actin, intermediate filaments, microtubules
(b) intermediate filaments, actin, microtubules
(c) microtubules, actin, intermediate filaments
(d) microtubules, intermediate filaments, actin
(e) intermediate filaments, microtubules, actin

32. Of the cytoskeletal elements listed below, which one is usually *not* involved in cell movement? _____

(a) actin
(b) microfilaments
(c) intermediate filaments
(d) microtubules
(e) myosin

33. Actin is a cytoskeletal protein, whereas histones are nuclear proteins that help package DNA. Thus, their functions are quite different. They do, however, share one quality, which is that _____.

(a) they both have a molecular weight greater than 100,000 daltons
(b) antibodies cannot be made for either
(c) they both co-purify on a gel filtration column
(d) they are both highly conserved in evolution
(e) they both have an ATPase fold requiring magnesium

34. Actin and microtubules differ from intermediate filaments in that _____.

(a) they form as polymers
(b) they are part of the cellular cytoskeleton
(c) they have plus and minus ends
(d) they are found in nerve cells
(e) none of the above

35. The term *"brush" border* is best matched with which of the following terms? _____
   (a) microtubules
   (b) microvilli
   (c) intermediate filaments
   (d) spectrin
   (e) myosin

36. Which of the following is *not* a commonly used tool for analyzing the assembly of G actin into F actin?_____
   (a) phase microscopy
   (b) viscometry
   (c) sedimentation
   (d) fluorescence spectroscopy
   (e) none of the above; that is, all are used for this type of analysis

37. Capping proteins _____.
   (a) inhibit the addition of actin monomers to F actin
   (b) bind to intermediate filaments, but not microtubules or actin filaments
   (c) are found in prokaryotes, but not eukaryotes
   (d) are related to **heat-shock proteins**
   (e) all of the above

38. Cytochalasin D _____.
   (a) is a fungal alkaloid
   (b) binds to the plus end of actin filaments
   (c) depolymerizes actin filaments
   (d) can stop cell movement in living cells
   (e) all the above

39. Phalloidin is often used in combination with a fluorescent dye for staining _____.
- (a) actin filaments
- (b) microtubules
- (c) intermediate filaments
- (d) myosin filaments
- (e) none of the above

40. Which of the following is a monomer? _____
- (a) F actin
- (b) $\alpha\beta$-tubulin
- (c) myosin I
- (d) myosin II
- (e) all the above

41. Which of the following types of myosin powers muscle contraction? _____
- (a) myosin I
- (b) myosin II
- (c) myosin V

42. There are several steps to the sliding filament hypothesis of muscle contraction. Which of the following occurs during the **power stroke**? _____
- (a) binding of ATP
- (b) release of ADP and P_i subsequent to the hydrolysis of ATP to ADP
- (c) binding of ADP
- (d) hydrolysis of GTP
- (e) binding of GDP

43. Which of the following actin structures is most closely associated with mitotic cells? _____
- (a) contractile ring
- (b) adhesion plaque
- (c) **adherens junction**
- (d) stress fibers

44. Cytoplasmic streaming is mediated by _____.
 (a) intermediate filaments
 (b) actin
 (c) myosin
 (d) both (b) and (c)
 (e) none of the above

45. Which of the following drugs does not belong with the rest? _____
 (a) colchicine
 (b) cytochalasin D
 (c) taxol
 (d) vinblastine
 (e) nocodazole

46. Colchicine is useful for karyotyping because _____.
 (a) it depolymerizes actin filaments only
 (b) it depolymerizes the mitotic spindle at high concentrations, thus blocking the cell in metaphase
 (c) it enhances the polymerization of the mitotic spindle
 (d) it blocks both actin filaments and microtubules
 (e) it inhibits chromosome duplication

47. If movement of materials is from the soma (cell body) of a neuron to the axonal terminal, it is referred to as _____.
 (a) synaptograde
 (b) lineargrade
 (c) **retrograde**
 (d) anterograde
 (e) none of the above

48. Kinesin is a motor protein that moves material _____.
 (a) from the minus end to the plus end of microtubules
 (b) from the plus end to the minus end of microtubules
 (c) both (a) and (b)
 (d) it is not known in which direction kinesin moves materials in cells

49. Radial spokes, nexin, and dynein are components most closely associated with _____.
   (a) centrioles
   (b) mitotic organizing centers
   (c) flagella
   (d) cilia
   (e) both (c) and (d)

50. How many complete and incomplete microtubules are present in cilia? _____
   (a) 5
   (b) 7
   (c) 9
   (d) 13
   (e) 11

51. The type of intermediate filament that is characteristically found in the cytoplasm of human skin cells is _____.
   (a) vimentin
   (b) desmin
   (c) keratins
   (d) lamins
   (e) neurofilaments

52. The type of intermediate filament that forms the karyoskeleton of the nucleus and serves, in part, to regulate the dissolution of the nucleus is _____.
   (a) vimentin
   (b) desmin
   (c) keratins
   (d) lamins
   (e) neurofilaments

53. The term *tetramers* is most closely associated with which of the following? _____
(a) actin filaments
(b) microtubules
(c) microfilaments
(d) intermediate filaments

54. Intermediate filaments are associated with which of the following junctions responsible for the cell-to-cell adhesion strength of **epithelial** cells? _____
(a) desmosome
(b) tight junction
(c) gap junction
(d) synapse
(e) zonula occludens

55. Epidermolysis bullosa simplex is a disease associated with defective intermediate filaments and is most commonly noted in which of the following tissues? _____
(a) liver
(b) pancreas
(c) skin
(d) intestine
(e) lung

Indicate whether the following statements are True or False.

56. Filopodia can be described as actin-containing sheets of membrane that extend in the direction of cell movement.

57. Adhesion plaques are points of attachment of cells on the ventral (substrate) surface of the cell.

58. Microtubules are most often congregated around the outer surface of cells in aggregated collections sometimes called *stress fibers*.

Self-Testing Questions

59. Immediately after the lag period of actin and microtubule assembly, an oligomer forms a nucleus that acts as an initial scaffold on which actin or tubulin units can bind, resulting in elongation of the polymer.

60. Microvilli contain actin filaments that affect their structure.

61. The "G" in *G actin* refers to "globular."

62. Actin filaments, microtubules, and intermediate filaments can all treadmill.

63. The **critical concentration** of actin is the concentration that results in no net growth of the tubule.

64. *Listeria monocytogenes* infects mammalian cells and is able to assemble a tail of intermediate filaments that it uses for propulsion.

65. Rigor mortis occurs because there is no ATP, causing the myosin filaments to bind strongly to actin filaments.

66. The **sarcoplasmic reticulum** stores and releases calcium during skeletal muscle contraction.

67. The T tubule is a special type of microtubule found exclusively in neuronal axons.

68. The basal body is associated with flagella and cilia.

69. In most cells the minus end of microtubules is near the cell's plasma membrane, whereas the plus end is closer to the center, near the **microtubule organizing center (MTOC)**.

70. Microtubules undergo covalent modifications such as addition of tyrosine and acetylation. These processes can be used to reveal the age of modified microtubules under study.

71. Microtubule assembly and disassembly occur preferentially at the plus end.

72. GTP-bound tubulin heterodimers at the tip of a microtubule are more resistant to depolymerization than GDP-bound tubulin heterodimers.

73. Kinesin and dynein are considered molecular motors.

74. Intermediate filaments are the tracks on which pigment molecules move in pigment cells.

75. AMP-PNP has been useful in showing that the hydrolysis of ATP is necessary for molecular motors to move vesicles on microtubular tracks.

76. Microtubules can be easily isolated and studied *in vitro*. Similarly, it is possible to examine how the molecular motors kinesin and dynein work by purifying them, tagging them appropriately, and then applying them to isolated microtubules. In a series of experiments, you find that the addition of AMP-PNP causes the kinesin to bind to the microtubules, but no movement occurs. Adding ATP, however, causes kinesin to move in the expected direction. What is the most likely reason for this result, and what information does it reveal about the importance of ATP and kinesin?

77. Cytochalasin treatment of polarized kidney cells *in vitro* causes a decrease in the total surface area of these cells. Explain this phenomenon.

78. The phosphorylation of lamins triggers the dissolution of the nuclear lamina which, in turn, results in the dissolution of the nuclear envelope during mitosis. Assuming that you have all necessary tools available, how would you do the following:
(a) show that lamin B is phosphorylated prior to **mitosis**;
(b) show that the protein kinase that phosphorylates lamin B is in the cytoplasm;
(c) determine when during the cell cycle this protein kinase is most active.

79. You are growing dorsal root ganglion cells in culture and treat the cells with a nonionic detergent and high salts.
(a) What would you expect to see left after this treatment, and how might you verify its composition?
(b) **Axons** (long processes emanating from the cell body and from whose distal tip neurotransmitter is usually released) of most neurons contain both microtubules and intermediate filaments. Some axons use microtubules as their main supporting structure, while other axons use intermediate filaments. How might you determine if dorsal root ganglion cells use microtubules as the major supporting cytoskeletal elements?

126 Topic 9 *Cytoskeleton and Cell Movement*

80. Normal human epidermal keratinocytes (NHEK) secrete laminin (note: this is lamin*in*, not lamin), an extracellular matrix protein. The ability of the cytoskeleton to regulate laminin secretion was investigated by adding either nocodazole, a microtubule inhibitor, or cytochalasin, an actin filament inhibitor, to NHEK cultures. Next, intracellular, ^{35}S-methionine-labeled laminin was immunoprecipitated and analyzed on an **SDS** gel autoradiogram. An idealized autoradiogram representing the data appears below. Do these data verify that actin filaments regulate laminin secretion by NHEK cells *in vitro*?

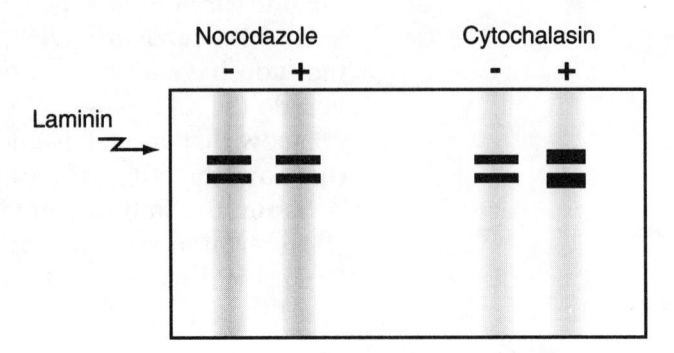

81. Colchicine and taxol have opposite effects, but both are used as anticancer drugs. Explain this apparent paradox.

82. Microtubules are dynamic structures that can transport vesicles and can shorten or lengthen, depending on the conditions. Alternatively, they can stay the same length. List three different mechanisms by which microtubules can maintain the same length.

83. Microtubules assembled *in vitro* with GMP-PNP are more stable than microtubules assembled *in vitro* with GTP. Explain why this is the case.

84. You isolate cilia in the laboratory and remove their outer membranes. When ATP is added, you expect to see the structure flex to one side. Instead, you notice that the organelle "telescopes." What might have gone wrong in the isolation process that could have accounted for this observation?

85. One of the best remedies for ingesting the highly toxic compound phalloidin is to eat raw meat. Why would this be true?

86. Many integral membrane proteins are free to move laterally in the plane of the plasma membrane, while others are restricted. Some of these restricted proteins are indirectly associated with actin filaments. Assuming you have a fluorescently tagged antibody that can bind to the extracellular **domain** of an integral membrane protein, how might you show that this Protein X, like **band 3** of the red blood cell membrane, is indirectly associated with actin filaments?

87. You investigate a novel organism that demonstrates cytoplasmic streaming. It appears very similar to **Nitella**, a green alga. Cytoplasmic streaming is fastest nearest the plasma membrane and slowest in the interior of the cell. You suspect that the streaming might be due to the interaction of actin and myosin. How might you use different approaches to show that this is the case?

88. How could you use isolated globular heads from myosin molecules to show that G actin adds faster to the plus end than it does to the minus end of an elongating F actin filament?

ATP

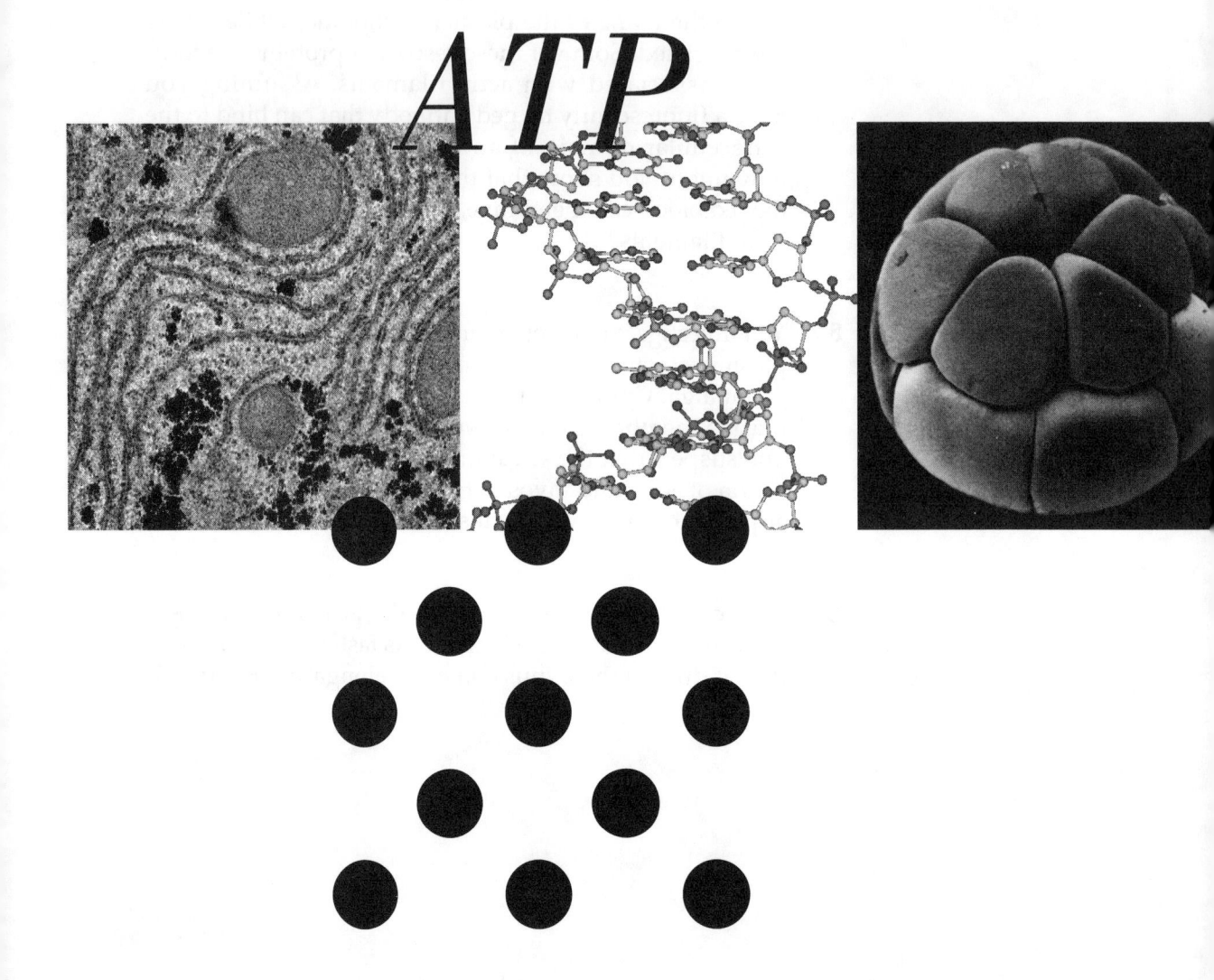

topic 10
Nervous and Immune Systems

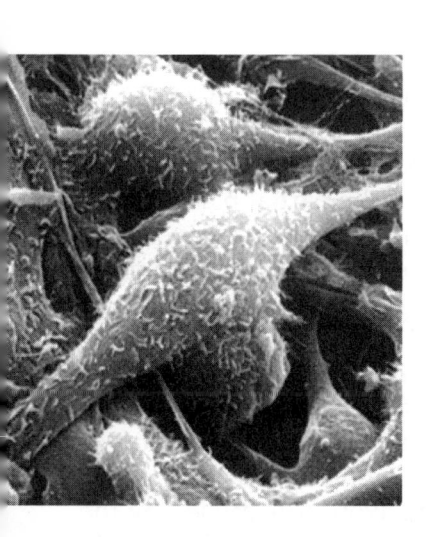

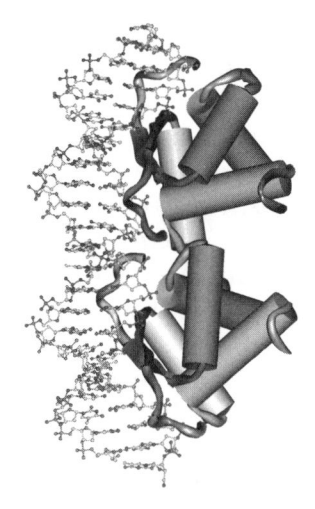

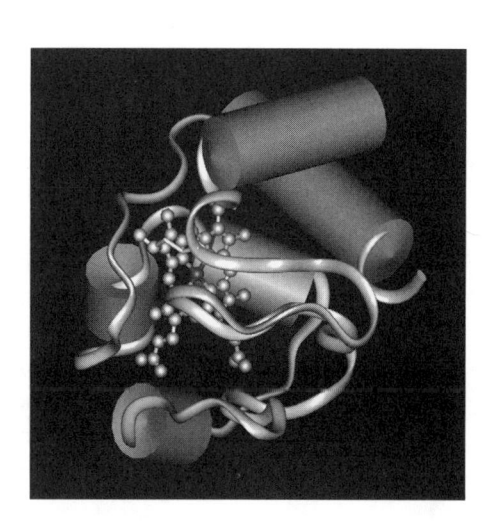

Summary

Neurons and immune cells are some of the most specialized animal cells. Both possess a "memory." Nerve cells are the basis of learning and memory; immune cells can recognize **antigens** that were introduced to an organism years prior to a subsequent exposure.

Neurons have many different shapes, but most have dendrites, a soma, and one or more axons. The dendrites of a neuron are short processes that emanate from the soma (cell body containing the nucleus). They usually receive incoming information from another neuron. The axon is usually much longer and transmits the electrical signal of a neuron to its terminus, where the information is transferred to an adjacent neuron. All this electrical activity is governed by resting potentials and **action potentials**.

The resting potential of a neuron can be calculated using the Nernst and Goldman equations. The typical resting potential of a vertebrate or invertebrate neuron is approximately -70 mV, with the inside negative. This potential is due primarily to the full permeability of K^+ as well as the partial permeability of Cl^- and Na^+. The Nernst equation can be used to predict the resting potential of a neuron if the neuron is permeable to only one ionic species, whereas the Goldman equation can be used to predict the resting potential of a nerve cell if the membrane is permeable to more than one type of ion. The action potential of a neuron is the signaling mechanism used by neurons. It is an abrupt **depolarization** of the neuron from -70 mV to $+55$ mV that occurs for only 2 milliseconds and passes from the cell body to the tip of the axon. It is caused by the inward flow of Na^+. Once the action potential reaches the tip of the axon, a depolarization occurs that releases a neurotransmitter. This neurotransmitter quickly diffuses across the neurochemical synapse, causing a depolarization in the adjacent dendrite of a different neuron. It should be noted that while neurochemical transmission is the major mode of nerve-to-nerve communication, direct electrical connection between two adjacent nerve cells can occur by gap junctions.

Summary

While much has been learned about the basis of single nerve cell function, much still has to be elucidated about the basis of learning and memory in multicellular animals. A number of model systems exist that are yielding significant clues. The *Drosophila*, or fruit fly, has been useful because many learning mutants have been developed that allow us to determine the molecular basis of learning. *Aplysia*, or the sea slug, has been studied by many researchers, most notably Eric Kandel and colleagues. *Aplysia* and its gill withdrawal reflex has allowed investigators to examine a simple behavior and define each neuron and its function underlying the entire behavior. Analysis of this model has led to theories about the molecular basis of habituation and facilitation, as well as models of learning and memory. Finally, study of hippocampal neurons in the mammalian brain has generated more ideas about learning and memory through the model of long-term potentiation, or LTP. LTP is based on the function of the glutaminergic synapse, whose repeated activity can cause an enhancement (i.e., potentiation) of further responses.

The immune system originated as a specialized network of cells that protects the body from invading pathogens such as viruses, bacteria, and other organisms. The immune system consists of two parts. The humoral immune system is governed by soluble antibodies that are produced by B cells in response to an invading organism or foreign molecule. The cellular immune system, on the other hand, is mediated by circulating cells such as T lymphocytes and macrophages.

Antibodies are the basis of the humoral response and are unusual molecules in many respects. They are "Y-shaped" molecules composed of both heavy and light chains. They are members of one of five different classes: IgM, IgD, IgG, IgE, and IgA. The antibody has one F_c region that doesn't vary within a class, and two F_{ab} regions that vary tremendously and can bind two antigens per molecule of antibody. Antibodies are generated through clonal selection, a process by which antigens can "select" antibodies on the surface of B cells. Some of these B cells end up being long-lived memory cells that can react to a recurrence of an invading organism; other B cells are short-term plasma cells that react immediately to a foreign antigen.

Self-Testing Questions

Match the item on the left to the most appropriate term on the right.

1. Glutamate
2. Acetylcholine
3. Goldman/Nernst
4. Dopamine
5. Opioid
6. Serotonin
7. Glycine
8. **NMDA**
9. Synapsin
10. Rutabaga

A. glutamate receptor implicated in long term potentiation

B. receptor can be nicotinic or muscarinic

C. derived from tyrosine

D. *Drosophila* learning mutant

E. excitatory amino acid

F. regulates synaptic vesicle release

G. equations useful for calculating resting potentials

H. a neuropeptide and analgesic

I. derived from tryptophan

J. inhibitory amino acid

Indicate whether the following statements are True or False.

11. The parasympathetic autonomic nervous system stimulates involuntary muscles and slows heart rate.

12. On-cell patch clamp technique can measure the current through a single channel on an isolated, detached patch of neuronal membrane.

Self-Testing Questions

13. The approximate transmembrane potential of most neurons is −20 mv, inside negative.

14. Presynaptic inhibition necessitates an axonal-axonal synapse.

15. The concentration of potassium is greater outside the neuron than inside the neuron.

16. One of the demyelinating diseases is myasthenia gravis.

17. The node of Ranvier is associated with non-myelinated neurons.

18. Acetylcholine is hydrolyzed by acetylcholinesterase.

19. Rab3 is associated with the neurotransmitter vesicle and is critical for docking of the neurotransmitter vesicle to the presynaptic membrane.

20. Neurotransmitter vesicles are often spontaneously released and result in a depolarization of approximately 20 mv.

21. Activation of the glycine or GABA receptors commonly results in a **hyperpolarization** of the postsynaptic cell.

22. The GABA receptor passes chloride ions down its gradient.

23. *Aplysia* is a sea slug that has been important in the elucidation of neural circuits.

24. The facilitator neuron of the *Aplysia* works through the release of dopamine and the activation of adenylate cyclase.

25. Long-term potentiation occurs in the hippocampus.

26. Macrophages are immune cells that can degrade foreign material.

27. Antibodies are produced by lymphokines.

28. An antigen is a **bivalent** molecule that can bind to foreign molecules.

29. Antibodies are bound to antigens by covalent bonds.

30. A clone is a group of cells all originating from the same ancestor.

31. The clonal selection theory has been disproved as the basis for antibody diversity.

32. During its antigen-dependent phase, a B cell secretes antibodies.

33. Memory B cells are activated during the antigen-independent phase of B cell maturation.

34. Severe combined immunodeficiency disease (SCID) is a group of diseases where one or many types of lymphocytes may be absent.

35. Myelomas are cancer cells that secrete large numbers of antibodies.

36. The hinge part of the antibody molecule is part of the L chain.

37. The Fab region of the antibody is also referred to as the constant region.

38. DNA rearrangement generates antibody diversity.

39. The somatic recombination hypothesis underlies antibody diversity.

40. Suppressor T cells act to down-regulate the immune system.

Fill in the blanks of the following stentences to make accurate statements.

41. Proteins are synthesized in dendrites and the cell body of neurons but not the _____.

42. The short processes that usually receive incoming information from presynaptic cells are referred to as _____.

43. Transport from the axon terminal to the nucleus is referred to as _____ transport, whereas transport from the nucleus to the axon terminal is considered _____ transport.

44. An all-or-none response is often called an _____.

45. A depolarization of a neuron can be defined as a change in transmembrane potential that moves _____ to 0 mv.

46. The type of potential change that occurs between the nodes of Ranvier is considered _____.

47. The type of electrical analysis most commonly used to record changes in membrane potential in cells is referred to as _____ recording.

48. Acetylcholine exposure can result in a depolarization in _____ muscle, whereas it can result in a hyperpolarization in _____ muscle.

49. The place on the axon terminal where neurotransmitter vesicles fuse is referred to as the _____ zone.

50. The type of glutamate receptor found on hippocampal cells that is blocked by magnesium is referred to as the _____ receptor.

51. *Drosophila* learning mutants, *rutabaga* and *dunce*, are defective in the _____ second messenger system.

52. The result of long-term potentiation is the enhanced release of _____ from the presynaptic cell.

53. Light causes channels in the rod photoreceptor to _____.

54. The key pigment molecule in rod photoreceptors is _____.

Self-Testing Questions

55. Transducin is activated by light-activated opsin and activates cGMP _____.

56. _____ is also called an antigenic determinant.

57. A _____ is a small substituent that, when attached to a molecule such as a protein, can generate an antibody response.

58. A _____ is that part of the poly-Ig receptor that is cleaved once an IgA molecule has been transported across an epithelial layer.

59. Antibodies are made by _____ lymphocytes.

60. The _____ theory has now been disproved and states that antibodies would fold in different manners once they encountered an antigen.

61. The _____ theory is now the accepted model of how antibodies can recognize nearly every possible antigen presented to the immune system.

62. _____ B cells are the lymphocytes that can remember that an antigen was introduced during the primary immune response.

63. T cells are so named because they mature in the _____ gland.

64. The ability of the immune system to recognize self from nonself is called _____.

65. Thy-1 is a cell surface marker present on _____ lymphocytes.

66. The **major histocompatibility complex (MHC)** was originally discovered in experiments involving _____ rejection.

67. Class I major histocompatibility complexes bind peptides derived from _____ proteins, whereas Class II major histocompatibility complexes bind peptides derived from _____ proteins.

68. Class II major histocompatibilty complexes are found mostly on _____.

69. Peptides degraded from foreign protein are complexed with the major histocompatibility complexes in the _____.

70. Patches form on B lymphocytes when a multivalent antigen is presented. This is due to the _____ nature of the antibodies on the cell surface.

Select the letter that best completes the following statements.

71. Which of the following best describes that part of the neuron that usually releases neurotransmitter? ____
 (a) soma
 (b) cell body
 (c) dendrites
 (d) axon terminal
 (e) none of the above

Self-Testing Questions

72. What is the significance underlying the axon hillock?

(a) it is usually immediately adjacent to the cell body
(b) it is characterized by voltage-gated sodium channels
(c) it initiates the action potential
(d) its function can be blocked by neurotoxins such as tetrodotoxin
(e) all of the above

73. A neuron that is the first to receive information from the environment is often called _____.

(a) an interneuron
(b) a sensory neuron
(c) a facilitating neuron
(d) a motor neuron
(e) a habituation neuron

74. An action potential is characterized by all of the following *except* it _____.

(a) is initiated by opening of voltage-gated potassium channels
(b) is regarded as a regenerative response
(c) is regarded as an all-or-none response
(d) does not degrade in magnitude with space or time
(e) is characteristic of transmembrane potential changes that occur in most axons

75. A chemical synapse is characterized by all of the following *except* _____.

(a) neurotransmitter is released
(b) synaptic cleft is present
(c) intracellular increase in calcium is critical for function
(d) membrane fusion and gap junction are evident
(e) a one millisecond delay in membrane potential change exists between presynaptic and post synaptic cells

76. The nicotinic acetylcholine receptor is best described as a _____.

(a) signal- (e.g., second messenger) activated receptor
(b) voltage-gated channel
(c) ligand-gated channel
(d) resting channel (one that is not governed by ligands, second messengers, or voltage)

77. Cobalt can block chemical synaptic function because it _____.

(a) binds irreversibly to sodium ions
(b) blocks the voltage-gated calcium channel in the axonal terminal
(c) directly inhibits synapsin
(d) directly blocks syntaxin
(e) activates axonal proteases and phospholipases

78. Which of the following messenger molecules is a gas and has been implicated as a retrograde messenger in long-term potentiation? _____

(a) cAMP
(b) cGMP
(c) calcium ions
(d) nitric oxide
(e) diacylglycerol

79. The resting potential in most neurons is primarily due to the permeability of _____.

(a) calcium
(b) chloride
(c) sodium
(d) potassium
(e) magnesium

Self-Testing Questions

80. The resting potential would be equal to E_K if it weren't for _____.

(a) the partial permeability of the plasma membrane to calcium ions
(b) the partial permeability of the plasma membrane to chloride ions
(c) the partial permeability of the plasma membrane to sodium ions
(d) both (b) and (c)
(e) the partial permeability of the plasma membrane to magnesium ions

81. Myelin _____.

(a) is Schwann cell membrane
(b) increases the speed of propagation of action potentials
(c) wraps around the axons
(d) increases the membrane resistance of axons
(e) all of the above

82. Graded potentials _____.

(a) are characteristic of sensory neurons
(b) occur in axons
(c) necessitate voltage-gated sodium channels
(d) are not affected by the diameter of the neuronal process
(e) all of the above

83. Action potentials are unidirectional because _____.

(a) sodium ions can only flow from outside to inside the cell
(b) the voltage-gated sodium channel has a transient period of inactivity during which it cannot be opened again
(c) calmodulin's activation by calcium is delayed
(d) potassium channels are slow to inactivate
(e) none of the above

84. Which of the following drugs can block potassium channels? _____
  (a) SOMAN
  (b) LSD
  (c) tetrodotoxin
  (d) tetraethylammonium
  (e) amphetamines

85. Postsynaptic responses can be considered either fast or slow. Slow responses are often coupled to _____.
  (a) microtubular assembly
  (b) G proteins
  (c) ligand gated channels
  (d) protein hydrolysis
  (e) gap junctions

86. The antibody's heavy chain consists of or contains _____.
  (a) part of the Fab fragment
  (b) part of the antigen binding region
  (c) part of the Fc fragment
  (d) part of the effector domain
  (e) all of the above

87. How many biding sites does a typical IgG molecule have? _____
  (a) 1
  (b) 2
  (c) 3
  (d) 4
  (e) 5

88. Which type of antibody is pentameric, i.e., consists of five, 4-chain antibody molecules? _____
  (a) IgM
  (b) IgD
  (c) IgG
  (d) IgE
  (e) IgA

89. Which type of antibody is the first class to be generated in response to an antigen? _____
   (a) IgM
   (b) IgD
   (c) IgG
   (d) IgE
   (e) IgA

90. Which type of antibody consists of several antibodies held together by the J chain? _____
   (a) IgM
   (b) IgD
   (c) IgG
   (d) IgA
   (e) both (a) and (d)

91. _____ is regarded as the major serum antibody, stimulating both complement and macrophages and having the ability to pass through the placenta to the fetus.
   (a) IgM
   (b) IgD
   (c) IgG
   (d) IgE
   (e) IgA

92. Which antibody is the least understood? _____
   (a) IgM
   (b) IgD
   (c) IgG
   (d) IgE
   (e) IgA

93. Which antibody is a monomer, dimer or trimer and linked together with a J chain? This type of antibody is transported through epithelial cells, and the receptor that does so is cleaved at the end of the process. _____
   (a) IgM
   (b) IgD
   (c) IgG
   (d) IgE
   (e) IgA

94. Which antibody is the one that complexes with an antigen and releases histamine from mast cells causing allergic reactions? _____
   (a) IgM
   (b) IgD
   (c) IgG
   (d) IgE
   (e) IgA

95. The antigen binding receptors in T cells are called _____.
   (a) T cell receptors
   (b) LDL receptors
   (c) IgA
   (d) IgM
   (e) none of the above

96. The type of T cell that kills cells that have foreign antigens on their surface is called _____.
   (a) helper T cell
   (b) suppressor T cell
   (c) killer T cell
   (d) cytotoxic T cell
   (e) both (c) and (d)

97. Cytotoxic T cells secrete molecules that induce cell death by _____.
- (a) necrosis
- (b) phagocytosis
- (c) apoptosis
- (d) all of the above
- (e) not known

98. Helper T cells _____.
- (a) secrete molecules that aid B cells
- (b) are also called T_H cells
- (c) can recognize foreign peptides on cell surfaces
- (d) are lost due to HIV infection
- (e) all of the above

99. A CD4+ cell describes _____.
- (a) a red blood cell
- (b) a helper T lymphocyte
- (c) an unstimulated B cell
- (d) a stimulated B cell
- (e) a cytotoxic T lymphocyte

100. Monocytes are mostly closely related to _____.
- (a) macrophages
- (b) B cells
- (c) helper T cells
- (d) cytotoxic T cells
- (e) suppressor T cells

101. The bone marrow is recognized as the primary organ for B lymphocytes because _____.
- (a) this is where B cells generate antibodies
- (b) antigens migrate to bone marrow
- (c) this is where B cells originate
- (d) this is where the secondary immune response occurs
- (e) this is where haptens reside

146 **Topic 10** *Nervous and Immune Systems*

102. Peyer's patches are peripheral lymphoid tissues located in ____.
   (a) the liver
   (b) the intestine
   (c) the pancreas
   (d) the smooth muscle lining the throat
   (e) the gall bladder

103. You are attempting patch clamp analysis for the first time and decide to examine the nicotinic acetylcholine receptor. You chose to examine the acetylcholine receptor on muscle cells from a donor with a previously uncharacterized muscular dystrophy.
   (a) You first perform the "on-cell" patch technique by adding acetylcholine to the preparation and find that, due to technical difficulties, you are unable to make a stable recording. Why might this be the case? (Hint: What does acetylcholine do to skeletal muscle?)
   (b) In the next series of experiments you decide to do an "outside-out" patch, but once accomplished, you are unsure if the resulting patch is "outside-out" or "inside-out." Using acetylcholine and assuming the receptor can conduct current, how might you determine if you have one or the other?
   (c) One of the possibilities underlying this neuromuscular disorder is that there might be fewer acetylcholine receptors than in nonafflicted individuals. How might you use the patch clamp technique to determine if this is the case?
   (d) You propose that the underlying reason for this muscular disease is that the acetylcholine receptor is less able to pass sodium-potassium to the same extent as channels from nonafflicted individuals. How might this be revealed in the electrical record doing "outside-out" patch clamp analysis?
   (e) Assuming that part (d) above is true, why might atropine (a plant compound that inhibits the breakdown of acetylcholine) or an atropine-like compound be considered as a possible treatment candidate for patients with this disease?

- (f) In later experiments you clone the gene coding for the acetylcholine receptor in afflicted individuals and find that there are lysine residues in the neck of the M2 helix where glutamic acid residues are found in nonafflicted individuals. How might this explain the patch clamp electrophysiology revealed in part d above?
- (g) There has been much talk about "gene therapy" for neurological disorders. How might the work you have just accomplished be translated into this type of clinical approach?

104. The surface/area ratio of neurons is quite different than other cells and influences the amount of ATP necessary to run the sodium/potassium pump. Explain this relationship.

105. Denervation can cause a re-localization of acetylcholine receptors. Explain how you might investigate this phenomenon using α-bungarotoxin, a snake toxin that binds tightly to nicotinic acetylcholine receptors.

106. Neuropeptides are a class of neuroactive compounds that are synthesized in the soma and then transported in vesicles to the axon terminal via microtubules. In a preparation of isolated neurons in culture, you suspect that a particular neuropeptide is located in a vesicle and that it is being transported via microtubular transport. The following questions relate to this experimental situation. Assume that you have antibodies to both the neuropeptide and tubulin and that you can radioactively tag anything that you choose.
- (a) How might you determine if this neuropeptide is, in fact, packaged in vesicles in contrast to localized diffusely through the cytoplasm?
- (b) When you add nocodazole to these isolated cells, the vesicle transport stops. How might you determine if

nocodazole is inhibiting vesicle transport via depolymerization of the microtubules?
(c) When cytochalasin B is added, vesicle transport still continues but the synaptic terminal retracts. How might you explain this phenomenon?
(d) When you analyze the release of the neuropeptides from the axonal terminal, you find that the removal of extracellular calcium inhibits the release. Why is this observation consistent with our notions of neurotransmitter release?

107. You are investigating a very unusual post-synaptic change in membrane potential in response to a current injection of the presynaptic cell. This change in membrane potential results in a depolarization from −70 mv to +55 mv that lasts for 1 millisecond, with a delayed repolarization event that lasts for 50 milliseconds.
(a) Draw an approximate electrophysiological trace based on the above description.
(b) Many extended repolarization phases such as the one described previously is due to delayed voltage-gated calcium channels in the post-synaptic membrane. Suggest experiments that would help determine if this is the case.
(c) An alternative possibility to part (b) might be the activation of a post-synaptic second messenger system. Explain this concept.

108. A neuron has the following ionic characteristics:

Ionic Species	Inside Conc. (mM)	Outside Conc. (mM)
W^{++}	0.02	2
X^+	10	250
Y^+	300	50
Z^{+++}	10	150
Organic Anions	150	-

(a) If these ions are the only major molecular species that influence the osmolarity of this neuron, is this neuron in osmotic balance with its surroundings?

(b) Calculate the Nernst values for ionic species W, X, Y and Z.
(c) What would be the resting membrane potential of this neuron if the resting permeabilities of W, X, Y, and Z were 0, 0.2, 1.0, and 0, respectively?

109. A myelinated mammalian neuron receives synaptic input from three different presynaptic processes labeled A–C as labeled below.

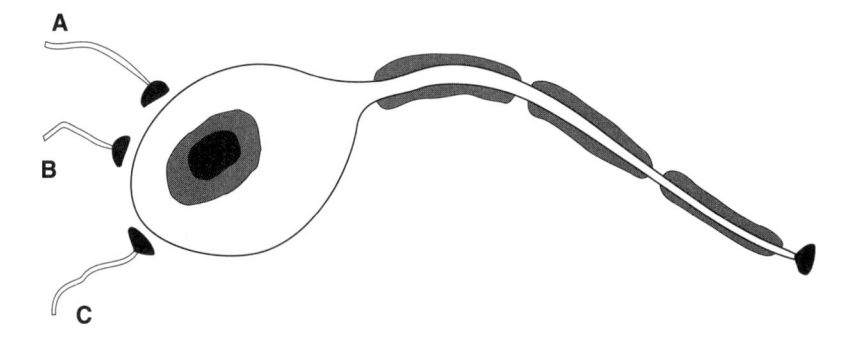

(a) Stimulation of C alone does not yield an action potential, but stimulation of A + C together does yield an action potential. How might this be explained?
(b) Simultaneous stimulation of A, B, and C together does not yield an action potential, yet as stated in part (a), stimulation of A + C together does result in an action potential. How might this be explained?
(c) Indicate the approximate location of voltage-gated sodium channels on the postsynaptic neuron.

110. You are doing a Western blot focusing on a poorly characterized protease (enzyme that can cleave proteins). You have been able to isolate the protease through ion exchange chromatography and gel filtration. A colleague in the laboratory has been able to make both a monoclonal and a polyclonal antibody against this protein. When you test both antibodies in the Western blot, you find to your great surprise that the monoclonal antibody yields one band, whereas the polyclonal antibody yields four bands! What is the simplest explanation for this apparent anomaly?

111. Tissue engineering is a new discipline of cell biology. It is the science whereby single cells are harvested from a donor and an engineered tissue is fabricated. Currently clinical trials are underway transplanting skin to burn victims as well as cartilage to individuals with severely damaged cartilage. A key part of this process is to harvest cells from the ultimate recipient to use to generate the new tissue. Why wouldn't a transplant from a different individual have equal likelihood of working?

112. The use of antibodies in purifying a protein from a heterogeneous mix of proteins is commonly used in the laboratory. In such a case you find that you are unable to "immunoprecipitate" protein Zed from a solution of mixed proteins. By diluting the mixture several fold, however, you can now successfully immunoprecipitate protein Zed. How might you explain this phenomenon?

113. You are interested in exploring the relationship between lymphocytes and the thymus gland. In so doing you purify lymphocytes and make a preparation of isolated thymus endothelial (blood vessel-lining) cells.
(a) What is the importance of the thymus gland in the immune system?
(b) When isolated thymus endothelial cells are placed in culture with lymphocytes the two cell types aggregate together. Why might this occur?
(c) How could you test the hypothesis presented in the answer to part (b)?

114. You are studying B cells and their response to applied antigen. You suspect that a particular antigen (Antigen Wop) is multivalent, that is, has many identical epitopes. How might you construct an experiment to prove that Wop is or is not multivalent?

115. A researcher attempts to generate an antibody against myelin membranes. Myelin membranes are produced by Schwann cells and oligodendrocytes and wrap around neurons. This electrical insulation confers speed to the action potentials that propagate down the axon. In order to accomplish this experiment, the researcher injects harvested myelin membrane into a rabbit. Several weeks later, another injection is given. During the course of this experiment the titre of antimyelin antibodies was measured and recorded as follows.

Weeks After Initial Injection	Relative Level of Antibodies
1	−
2	+
3	++
4	++
5	++
6	++++++++
7	++++++++
8	++++++++

During weeks 7 and 8 the animal generates tremors and has difficulty walking.

(a) At what approximate time was the second injection given?

(b) Why is there such a large increase in circulating antibodies starting at week 6?

(c) What might have accounted for the tremors seen in the rabbit?

116. You infect mice with a virus that invades epithelial cells and kills them. Your intent is to study the response of the cytotoxic T cells to virally infected cells. You plan to separate T cells from other blood cells so that they can be put in culture with the infected epithelial cells.

(a) How might you separate T cells from B lymphocytes and other blood cells?

(b) You find that when cytotoxic T cells are placed in culture with the virally infected epithelial cells, the epithelial cells shrink and the nucleus appears to have several large clumped areas. How might this be explained?

(c) New virally infected epithelial cells were then placed in the cell culture medium (no cells present) from the previous experiment and the epithelial cells died. What can be inferred from this experiment?

(d) If the virally infected epithelial cells are pretreated with an antibody directed against the major histocompatablity complex prior to presentation to the cytotoxic T cells, the cytotoxic T cells are unable to kill the epithelial cells. Why is this the case?

(e) When virally infected cells from a different mouse replaced the virally infected mouse cells from the original experiment, these epithelial cells did not die. What is the basis for this observation?

ATP

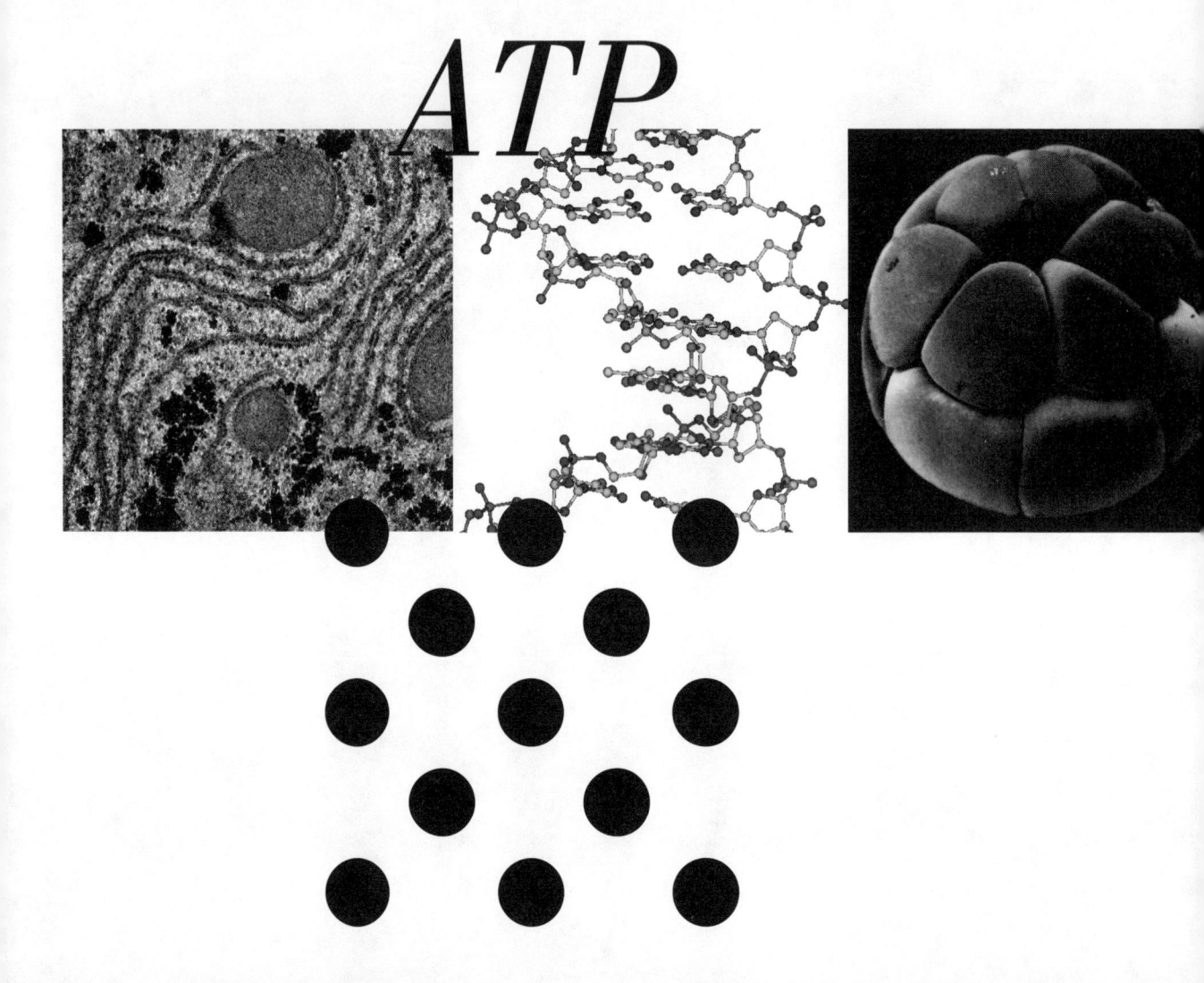

topic 11
Extracellular Matrix and Cell-Cell Interactions

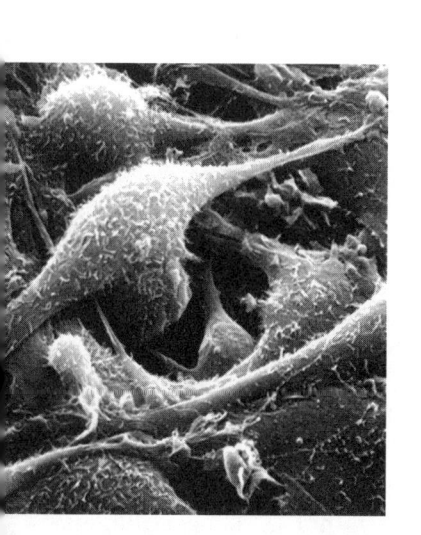

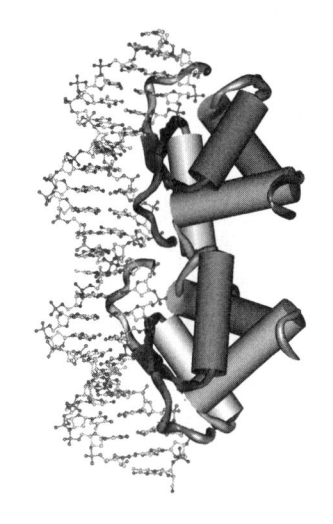

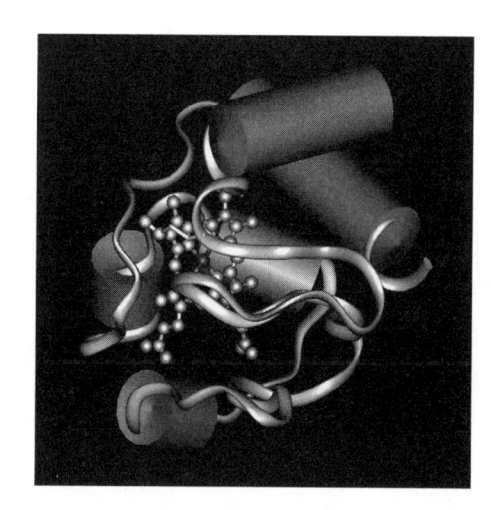

Summary

The extracellular matrix is a diverse collection of molecules secreted by cells. It has a wide-ranging impact on cell and tissue behavior. The extracellular matrix of animals can be divided into loose and dense connective tissue. Loose connective tissue forms the matrix on which and in which cells reside. It is a very porous and hydrated environment that consists, in part, of collagen, the major structural element of the extracellular matrix, and hyaluronan, a large and hydrated molecule. Dense connective tissue includes tendon, cartilage, and bone. The dense matrices are less hydrated than those of loose connective tissue and are important for strength and flexibility. As stated previously, collagen is the major extracellular matrix molecule.

Collagen is the most abundant animal protein and exists in at least 14 different varieties. Nearly all types are composed of three chains woven tightly together in a triple helix. Collagen is thus very fibrous and not soluble in water. Type I collagen is present in skin, tendon, and bone, whereas type IV collagen is the predominating extracellular matrix molecule present in all **basal laminas**, or basement membranes. Collagen is rich in proline and lysine, and every third amino acid is a glycine. The glycine is critical because its small R group allows for the twist of the triple helix. Collagen is synthesized and secreted as tropocollagen, with **propeptides** at both ends that direct the extracellular assembly of the molecule into collagen fibrils. The collagen fibrils are covalently linked in the extracellular environment by lysine oxidase so that many have a 67-nm staggered periodicity. Vitamin C is important in the synthesis of collagen because, if this cofactor is not present, collagen will not be properly hydroxylated in the endoplasmic reticulum and will fail to organize properly, leading to the deterioration of blood vessels. The extracellular matrix of animal cells also consists of other molecular species.

Hyaluronan is a very long molecule that can be measured in micrometers! It is a negatively charged polysaccharide that forms hydrated gels and gives tissue a porous and spongy property. It consists of 50,000 repeats of the simple disaccharide glucronic acid and N-acetylglucosamine. Owing to the number of hydrophilic groups on this immense molecule, hyaluronan will bind up to 1000 times its space in water. The hydrated nature facilitates cell migration. The extracellular matrix also consists of various **glycosaminoglycans**, such as chondroitin sulfate, dermatan sulfate, and heparan sulfate, among others. Other extracellular matrix molecules can link molecules together. Laminin, for instance, is a cross-shaped molecule found in the basal laminas underlying epithelial cells. It has multiple binding sites for receptors, collagen, and heparan sulfate **proteoglycans**. Laminin can also bind to integrin receptors that often link cells to the underlying substrate. Fibronectin is another example of an extracellular matrix linker molecule that binds many diverse molecular species including DNA, collagen, integrin receptors, and heparan sulfate. Fibronectin is important in the wound-healing process because it promotes cell migration. The intricate weaving of these diverse molecules that confer both a hydrated nature and an interlocking meshwork makes the extracellular matrix one of the most interesting structures in living systems.

Cells associate with one another using a variety of specific mechanisms. Cell adhesion can occur through cell junctions, a topic discussed in the cell membrane topic. While cell junctions are very important to cell-to-cell adhesion, they often don't confer the cell-to-cell specificity that cell adhesion molecules can impart to adjoining cells. Cell adhesion molecules can be sorted according to their dependence or lack thereof on calcium as a cofactor to effect cell adhesion. E-cadherin, for instance, is a calcium-dependent cell adhesion molecule predominantly found in epithelial cells. N-CAM is a member of the Ig superfamily and is a calcium-independent cell adhesion molecule. It is important for the migration of neurons during development.

In summary, the coordinated effort of the extracellular matrix molecules and cell adhesion molecules allows cells to maintain a differentiated state, remain in place, and recognize and communicate with cells of the same type.

Self-Testing Questions

Match the molecule on the left with the appropriate description on the right.

1. E-cadherin
2. N-CAM
3. Laminin
4. Fibronectin
5. Integrin
6. Collagen
7. Hyaluronan
8. Nidogen
9. Lysyl oxidase
10. Cellulose

A. Cell membrane receptor that links to collagen, laminin, fibronectin

B. Cell adhesion molecule present on neurons that changes its molecular weight during development

C. Enzyme that cross-links collagen molecules

D. Triple helix molecule present in loose connective tissue

E. Calcium-dependent cell adhesion molecule present in **epithelium**

F. Major structural protein of basal lamina—has 3 globular domains

G. Major structural protein of basal lamina—is cross-like molecule with multiple binding sites

H. Plant cell extracellular matrix component

I. Dimeric molecule that has RGD sequence that binds to integrin receptors

J. Large, hydrated molecule that lubricates joints

Self-Testing Questions

Indicate whether the following statements are True or False.

11. P-cadherin is concentrated in desmosomes.

12. The cadherins are calcium-dependent cell adhesion molecules.

13. N-CAM is a member of the Ig superfamily.

14. N-CAM is a calcium-dependent cell adhesion molecule.

15. Each type of N-CAM is coded by a separate gene.

16. Glycosaminoglycans are polysaccharide chains of repeating disaccharide units.

17. Glycosaminoglycans are very strongly hydrophobic.

18. Mature collagen fibers have a characteristic banding pattern of 67 nm that is due to the periodic and staggered arrangement of collagen molecules.

19. Collagen helices are held together primarily by hydrogen bonds.

20. Collagen molecules are linked together to form fibrils due to the cross links of alanine side groups.

21. The major function of collagen is to resist tensile forces.

22. The RGD sequence is an amino acid sequence common to fibronectin that causes one of the fibronectin dimers to bind to the other fibronectin dimer.

23. Multiple forms of fibronectin are produced by **alternative RNA splicing**.

24. Basal laminae consist of two layers, the clear lamina densa and the more darkly staining, lamina lucida.

25. Laminin has multiple binding sites for collagen and nerve cells.

26. Integrins need to bind manganese to function properly.

27. Integrins are dimeric molecules that are covalently linked together.

28. Many integrin receptors bind to actin filaments within cells and thus link the internal cytoskeleton of the cell to the extracellular matrix.

29. **Cancer** cells often synthesize more fibronectin than normal, non-transformed cells.

30. Integrins are homophilic-like cadherins.

Select the letter that best completes the following statements.

31. Tissues are often dissociated into single cells by treating them with EDTA. Why is EDTA effective in this manner? ____

(a) EDTA is a phospholipase
(b) EDTA is a phospholipase inhibitor
(c) EDTA is a calcium chelator
(d) EDTA is a protease
(e) EDTA is a **nuclease**

32. The main adhesion molecule that holds embryonic cells together is ____.
(a) laminin
(b) fibronectin
(c) cadherins
(d) N-CAM
(e) collagen

33. Uvomorulin is an alternative name for ____.
(a) E-cadherin
(b) P-cadherin
(c) N-cadherin
(d) N-CAM
(e) aggrecan

34. Which of the following is considered homophilic? ____
(a) integrin receptors
(b) cadherin
(c) laminin
(d) proteoglycans
(e) EGF receptors

35. Selectins mediate cell-cell adhesion in ____.
   (a) blood-capillary **endothelial** cells
   (b) the liver
   (c) the pancreas
   (d) bone
   (e) all of the above

36. Loose connective tissues are characterized by which of the following? ____
   (a) hydration
   (b) sparse number of cells
   (c) lots of collagen
   (d) fibroblasts
   (e) all of the above

37. Which of the following is *not* regarded as an extracellular matrix component? ____
   (a) laminin
   (b) proteoglycans
   (c) intermediate filaments
   (d) Type I collagen
   (e) Type II collagen

38. The basal lamina can often contain all of the following except ____.
   (a) Type IV collagen
   (b) Type I collagen
   (c) laminin
   (d) fibronectin
   (e) entactin

39. The term "ground substance" is often used to refer to ____.
   (a) chondrocytes
   (b) glycosaminoglycans and proteoglycans
   (c) laminin
   (d) fibronectin
   (e) fibroblasts

40. Osteoblasts are best matched with ____.
- (a) cartilage
- (b) basal lamina
- (c) hyaluronan
- (d) bone
- (e) epithelial cells

41. Which of the following is *not* a member of the glycosaminoglycan family? ____
- (a) elastin
- (b) keratan sulfate
- (c) heparan sulfate
- (d) dermatan sulfate
- (e) hyaluronan

42. Of the molecules listed, which has the largest molecular weight? ____
- (a) glycogen
- (b) collagen molecule
- (c) laminin
- (d) fibronectin
- (e) hyaluronan

43. "Resisting compressive forces" is best matched with ____.
- (a) collagen
- (b) laminin
- (c) integrins
- (d) hyaluronan
- (e) fibronectin

44. Aggrecan is a major component of ____.
- (a) the basal lamina
- (b) cartilage
- (c) bone
- (d) blood cells
- (e) the internal cytoskeleton of most cells

45. Proteoglycans can bind _____.
   (a) water
   (b) growth hormones
   (c) proteins
   (d) proteases
   (e) all of the above

46. Which of the following is partially assembled in the extracellular matrix? _____
   (a) laminin
   (b) collagen
   (c) integrins
   (d) fibronectin
   (e) all of the above

47. Collagen has which of the following as a repeating amino acid? _____
   (a) alanine
   (b) tyrosine
   (c) phenylalanine
   (d) glycine
   (e) none of the above

48. Which of the following is in the correct order from smallest to largest? _____
   (a) collagen molecule, collagen fibril, collagen fiber
   (b) collagen fiber, collagen fibril, collagen molecule
   (c) collagen fibril, collagen molecule, collagen fiber
   (d) collagen fibril, collagen fiber, collagen molecule
   (e) collagen molecule, collagen fiber, collagen fibril

49. Which of the following collagen subtypes can self-assemble into a sheet-like array? _____
   (a) Type I collagen
   (b) Type II collagen
   (c) Type III collagen
   (d) Type IV collagen
   (e) Type V collagen

50. A molecule that could be described as a "rubber band" is _____.
   (a) integrin
   (b) elastin
   (c) nidogen
   (d) laminin
   (e) fibronectin

51. Fibronectin _____.
   (a) is a linker molecule
   (b) is a protein
   (c) is an extracellular adhesive protein
   (d) has multifunctional binding domains
   (e) all of the above

52. Basal laminae _____.
   (a) are filters
   (b) can direct cell polarity
   (c) induce cell differentiation
   (d) promote cell migration
   (e) all of the above

53. The plant cell wall is considered an extracellular matrix structure. It is composed of _____.
   (a) cellulose
   (b) hemicellulose
   (c) pectin
   (d) all of the above
   (e) none of the above

54. Many of the extracellular matrix components such as cellulose are deposited in parallel strands. This polarity is directed, in part, by _____.
   (a) the Golgi apparatus
   (b) microtubules
   (c) actin filaments
   (d) collagen fibers
   (e) intermediate filaments

55. Which of the following cadherin molecules is located primarily on the placenta and other embryonic tissues? _____
 (a) E-cadherin
 (b) P-cadherin
 (c) N-cadherin
 (d) L-cadherin
 (e) S-cadherin

56. Collagen is the most abundant animal protein in the world. Human disease states exist that are characteristic of aberrant collagen synthesis/secretion. Explain how the following might result in collagen defects.
 (a) Defective lysyl oxidase.
 (b) Lack of vitamin C in the diet.
 (c) Defective C and/or N-terminal propeptide.

57. The lack of vitamin C affects collagen synthesis due to reasons that you described in the answer to question 56. In the last century the vitamin C-deficiency disease, scurvy, plagued sailors and caused blood vessels to become loose and fragile due to deficient collagen synthesis. This was because defective collagen would be degraded prior to secretion. Yet, remarkably, other collagen-rich tissues such as the cornea and the dermis of the skin were not similarly affected. What does this imply about the turnover of collagen in these tissues?

58. A newly isolated epithelial cell line expresses X-cadherin and normally aggregates with other like epithelial cells in culture. Yet this cell line does not adhere to a neuronal cell line when the two types are mixed in culture. How could you use both these cell lines and any molecular biology technique of your choice to determine if X-cadherin promotes cell adhesion through a homophilic interaction? (Assume that the only manner in which the epithelial cells adhere to each other is through the X-cadherin molecule.)

59. Many cells link to the extracellular matrix via fibronectin. Fibronectin is often called a linker molecule because it links integrin receptors in the plasma membrane to the basal lamina. The RGD sequence in the fibronectin molecule appears to be a key recognition peptide for the integrin receptor. How could you use cultured epithelial cells containing integrin receptors, fibronectin, and the synthetic peptide RGD to show that this tripeptide sequence is critical for integrin-fibronectin attachment?

60. You are investigating N-CAM in the developing brain. You have an antibody that recognizes N-CAM and you track the protein using Western Blots and SDS polyacrylamide gel electrophoresis **(SDS)** that separates based on molecular weight. You notice that a Western blot of this cell adhesion protein appears different when N-CAM is viewed in neonatal brain tissue compared to adult brain tissue. The data appear below.

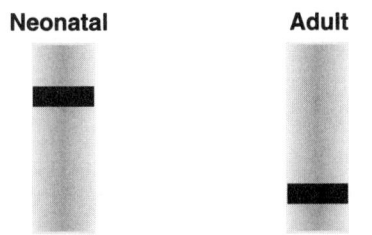

Neonatal   **Adult**

How might you explain the apparent difference and significance between these two blots?

61. You discover a new type of collagen that has the same number of amino acids per helix but a very different amino acid sequence than the conventional collagens. You suspect that this new collagen might have fewer hydrogen bonds between the chains and thus be less stable than well-known collagen subtypes. How might you demonstrate this using heat and a spectrophotometer?

62. Human keratinocytes (skin epithelial cells) can be grown in culture and under the correct conditions form a multi-layered epidermis. Cells are initially seeded on a microporous membrane (filter-like cell culture device) that has been pre-coated with a Type I collagen matrix (see Panel A below). The medium needs to contain low levels of extracellular calcium to encourage the cells to divide and move across the membrane. Once the cells have covered the entire surface as a monolayer, the extracellular calcium is elevated to 1.5 mM and the cells begin to stratify (see Panel B). Once they have stratified, the culture is placed at the air-liquid interface—a necessary step to complete differentiation (see Panel C). The type of collagen matrix form used is important. If a dry collagen film is placed on the microporous membrane, stratified human epidermis will not result. If a hydrated Type I collagen gel is used instead, an epidermis will be formed that secretes its own basal lamina as shown in the figure.

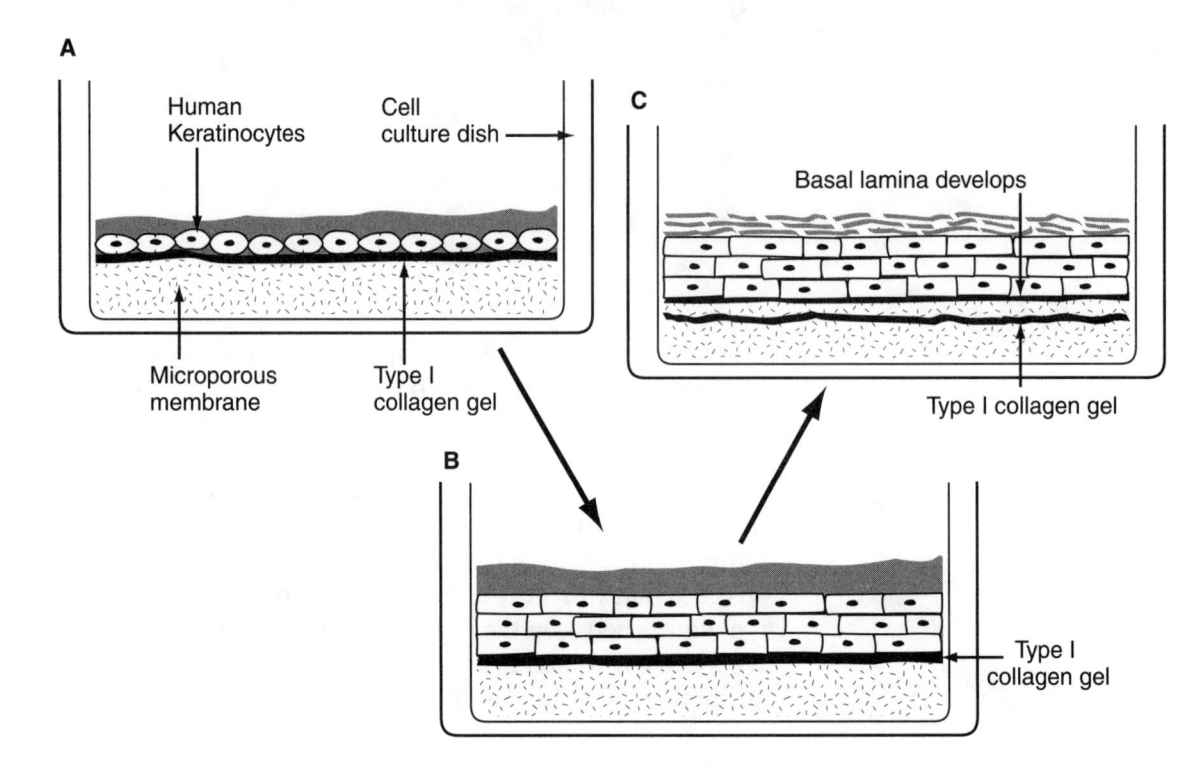

A commercially available form of this "engineered" human epidermis is being used as an animal alternative for product safety testing by many cosmetic and pharmaceutical industries.

(a) Why is it important that low levels of calcium be maintained in the early stage of developing this engineered human epidermis?

(b) As stated in the question, the matrix form of the gel is critical to the final development of this engineered human epidermis. It is well known that human keratinocytes are highly dependent on growth factors for growth and differentiation. Given that fact, why might the addition of proteoglycans to the Type I collagen matrix be an important consideration?

(c) Fully developed engineered human epidermal models will develop a basal lamina immediately under the epithelial sheet, as shown in Panel C on page 168. Electron micrographs of this basement membrane show a very thin sheet that is difficult to discern due to its thin nature. How might you convince someone that this basal lamina is, indeed, a functional basal lamina?

(d) It is possible that the sheet-like basal lamina is somehow originating from re-aggregation of collagen fibers from the Type I collagen and thus is not a true basal lamina. How might you construct an experiment to show that this basal lamina is actually being synthesized and secreted by the overlying keratinocytes?

(e) How might you demonstrate that the *in vitro* epidermis is histologically and biochemically similar to *in vivo* epidermis?

(f) One of the biggest challenges in constructing an *in vitro* epidermis is to ensure that the barrier property of this engineered tissue is similar to *in vivo* epidermis. One analytical method used to compare the barrier properties is to add ^3H_2O to the apical compartment of the human epidermal model detailed in Panel C and then determine how much of this radioactive tracer percolates to the basal compartment over time. Data from such an experiment are as follows.

	In vitro Epidermis	*In vivo* Epidermis
Total CPM placed in apical compartment	2,089,347	1,999,307
CPM in basal compartment after 1 h.r	2,457	1,209
CPM in basal compartment after 10 hr.	23,703	11,498
CPM in basal compartment after 20 hr.	45,098	23,478

CPM = counts per minute from scintillation counter

What can you conclude from this experiment?

(g) Many cancer cells can **metastasize** to other tissues by penetrating the basal lamina. Microporous membranes can be used to study the ability of cancer cells to migrate across a basal lamina. In one such model, the renal cell line, (MDCK) is grown on microporous membranes. After seven days these cells secrete a basal lamina as shown below. Next, the MDCK cells are

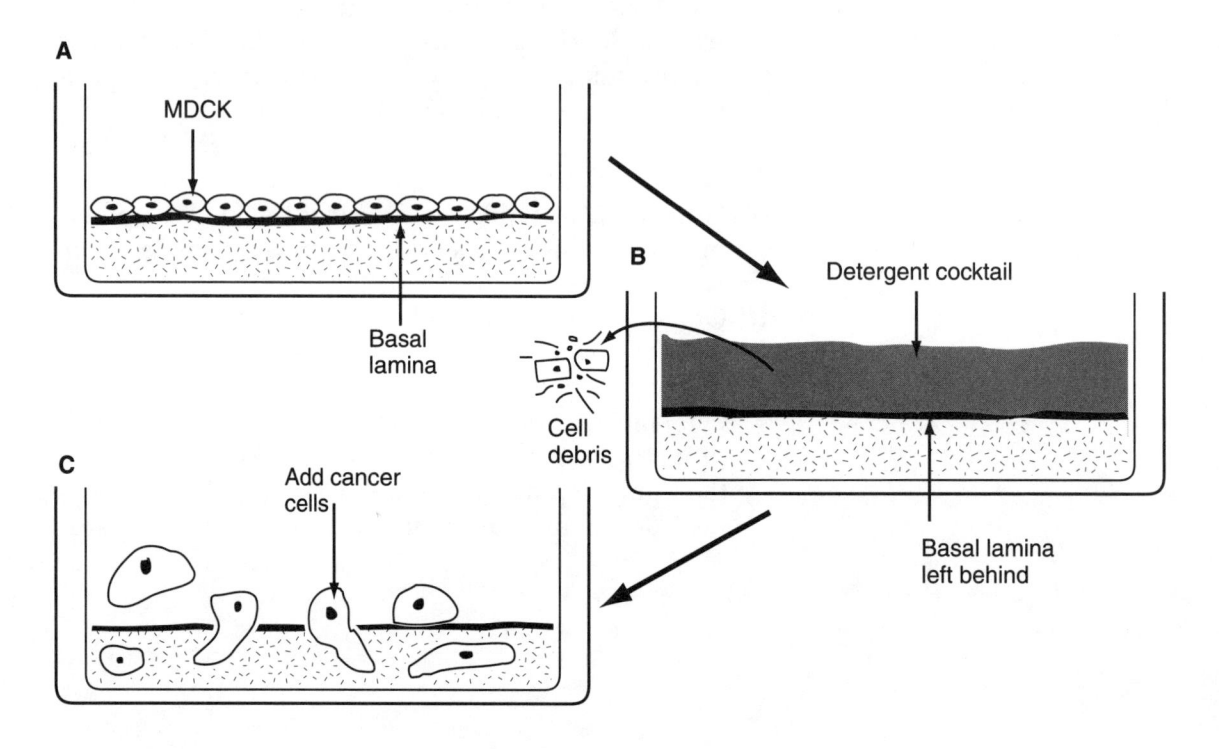

stripped off the membrane using a detergent cocktail leaving the basal lamina intact. Finally, cancer cells are applied to the apical surface and their ability to migrate through the membrane to the basal side is quantified.
   (i) What mechanism might these cancer cells use to accomplish this task and how might you prove or disprove your hypothesis?
   (ii) These cancer cells synthesize and secrete less fibronectin than their normal, nontransformed counterpart. Why is this fact not surprising?

63. Cells bind to each other via multiple mechanisms. Some cell adhesion molecules are heterophilic while others, such as the cadherins, are homophilic. In rare instances, however, cells need an extracellular linker molecule that links cell adhesion molecules together. In a cell culture system you find that Cell Z aggregates but cannot be easily dissociated with the calcium chelator, EDTA. You suspect that Cell Z has cell adhesion molecules but also uses an extracellular linker molecule that is secreted by these cells *in vitro* as shown below.

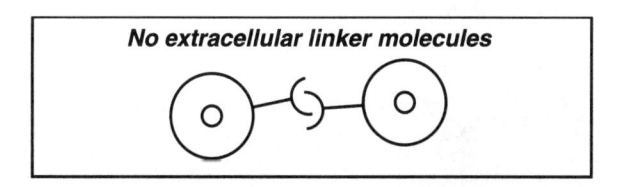

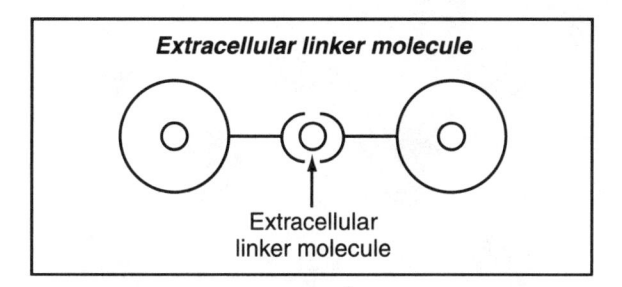

(a) How could you demonstrate that cell adhesion molecules are involved in the aggregation of this cell type?
(b) How could you show that extracellular linker molecules, secreted by Cell Z, are operating in this system?

ATP

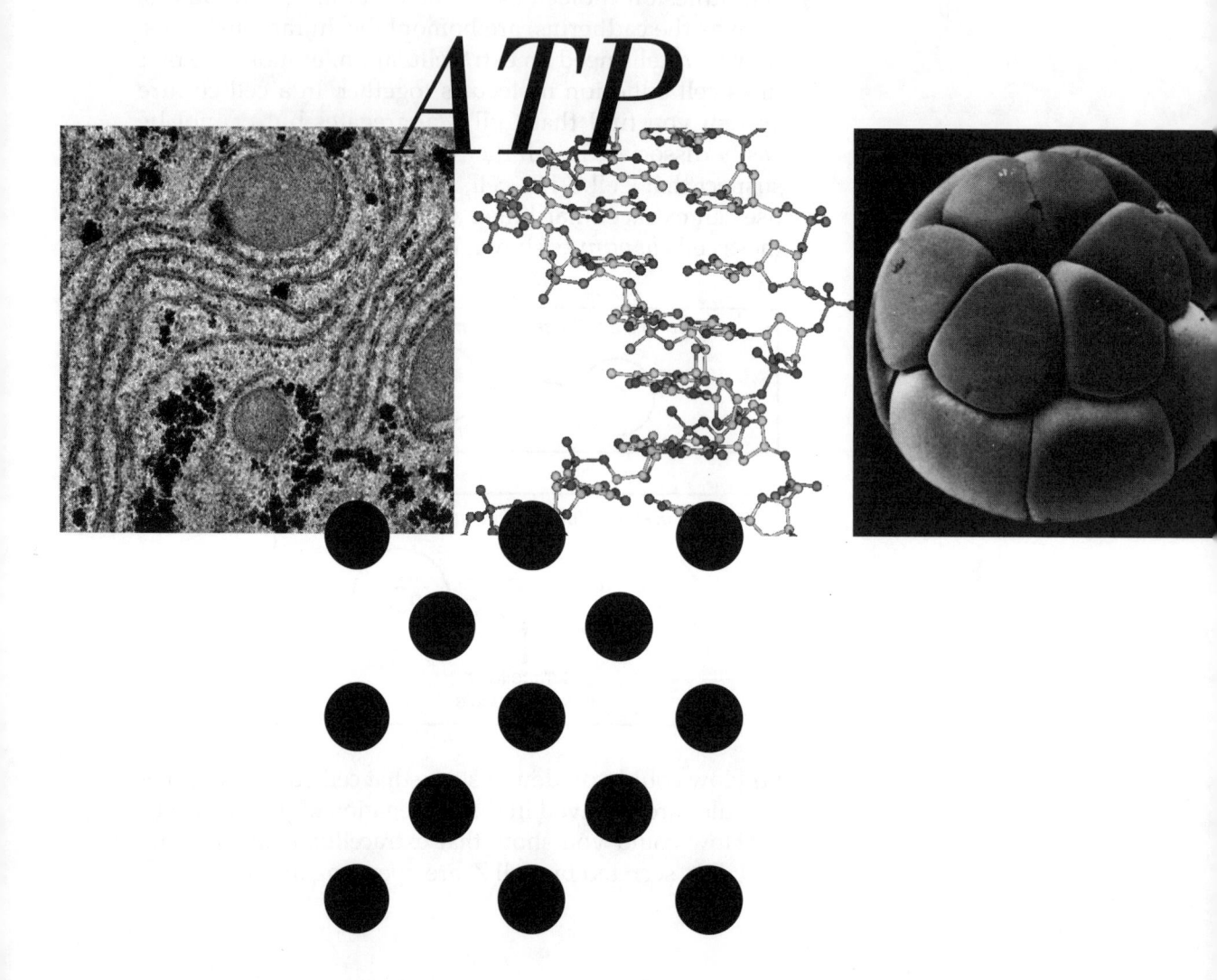

topic 12
The Cell Cycle and DNA Repair

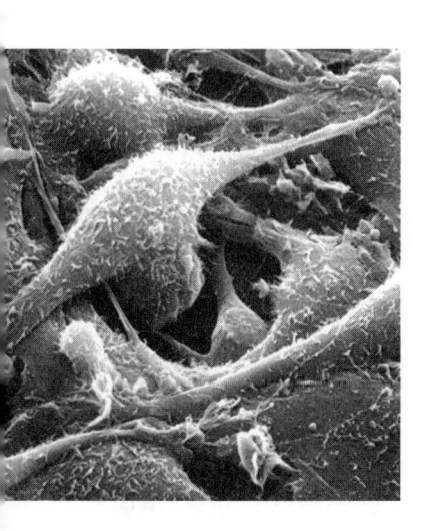

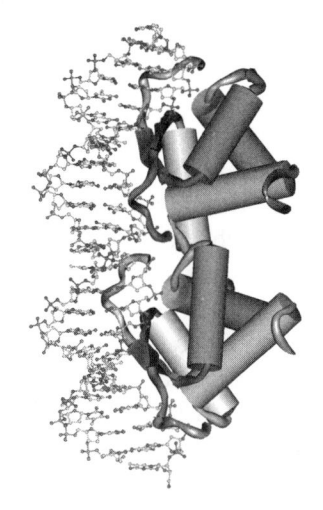

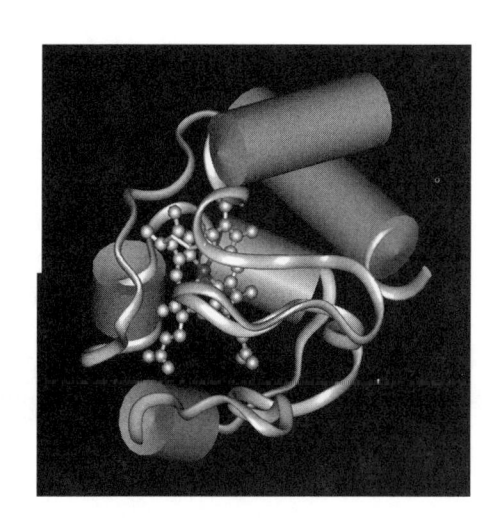

Summary

All cells undergo a cell division cycle during which specific metabolic activities occur preparing the cell to split into two daughter cells. The **cell cycle** has been very well defined in eukaryotic cells and is divided into a number of cell cycle compartments. G_0 is that part of the cycle during which no cell division activities are under way. A cell in G_0 is regarded as **differentiated** and is often carrying out some type of function specific to that cell type, such as insulin secretion by pancreatic cells. When cells are triggered to enter the cell division cycle, they first enter G_1, during which proteins critical to the cell division process are synthesized. Once cells have moved through G_1, they enter DNA synthesis, or the S phase. During this phase cells replicate their DNA. Cells then enter the G_2 compartment, during which more protein synthesis occurs. Finally the cells enter the most dramatic phase of the cell cycle, mitosis, or M. Mitosis is characterized by **karyokinesis**, the separation of the nucleus into two daughter nuclei, and **cytokinesis**, the microtubule and actin-mediated separation of the cytoplasm to form two new daughter cells. Mitosis can also be divided into prophase, metaphase, anaphase, and telophase according to the position of the **chromatids** and the degree of separation of the two new daughter cells.

The factors that control cell division have been studied actively in the past several years. Through two important independent lines of investigation, an M-phase promoting factor (or maturation-promoting factor) was discovered that triggers mitosis. This MPF leads to the dissolution of the nuclear envelope during mitosis. Further analysis revealed that MPF works through cyclins and a cyclin-dependent protein kinase. Briefly, cyclin B increases just prior to mitosis and combines with a cyclin-dependent protein kinase, forming an active MPF complex. This MPF phosphorylates lamin, among other substrates, and causes the dissolution of the nuclear envelope, triggering the initial phases of mitosis. At the end of mitotic prophase, polyubiquitination of the mitotic cyclin causes activation of the destruction box contained within the cyclin B molecule, causing its degradation. This decreases MPF activity to low levels until cyclin B is synthesized once again just prior to the next round of mitosis. Further analysis has shown that eukaryotic cells have multiple cyclins and cyclin-dependent protein kinases that rise and fall in activity during different phases of the cell cycle.

During the cell cycle it is critical that processes be in place to ensure that cell division occurs with near perfect fidelity. There are several checkpoint controls in the eukaryotic cells. One checkpoint in G_1 detects DNA damage, and if mismatched bases are indeed present, the cell cycle stops in G_1 and sometimes the cell commits suicide through **apoptosis**. One important G_1 checkpoint regulator protein is p53. People with deficient p53 are unable to block the replication of damaged DNA, and as a result mutations and cancer can occur. Other checkpoints occur in G_2 that can detect unreplicated DNA and DNA damage. A final checkpoint occurs in mitosis, where if the mitotic spindle is not properly aligned, the cell cycle arrests in M. The microtubular destablizing agent colchicine arrests cells in M for this reason.

DNA mispairing is somewhat of a paradox. While mismatching can lead to cancer if not detected by one of the checkpoint controls described earlier, mispaired bases are the basis of evolution. DNA mutation can occur through a number of mechanisms that include spontaneous mutation as well as radiation and chemically induced mutation. The types of DNA damage that occur include depurination, deamination, thymine dimerization, and production of reactive metabolites such as free radicals. There are several different enzymes that can repair DNA mutations. They include DNA polymerase, AP **endonuclease**, and DNA glycosylases. Inherited defects in repair enzymes can result in human disease states such as Xeroderma Pigmentosum, a malady characterized by increased sensitivity to ultraviolet light.

In summary, a combination of internal cell cycle checkpoints coupled with a variety of DNA repair enzymes ensures that cell division occurs with the best possible fidelity. When one or more of these enzymes are either absent or not functioning properly, the chance of having a normal cell tranform to a cancer cell is increased.

Self-Testing Questions

Match the phase of the cell cycle on the left with its appropriate description on the right.

1. G_2
2. G_0
3. S
4. M
5. G_1

A. Compartment during which chromosomes separate from each other and two new daughter cells form

B. Period immediately following synthesis (S)

C. Exit point from the cell cycle, during which a cell is not participating in any cell division activity

D. First part of the active cell cycle, during which protein synthesis is critical

E. Compartment during which DNA is synthesized

Indicate whether the following statements are True or False.

6. Chromosomes in mitosis are attached to the spindle microtubules at the centromere.

7. Cytokinesis refers to the physical separation of two new daughter cells during mitosis.

8. **Interphase** is also referred to as mitosis.

9. Fusion of S cells with G_1 cells cause G_1 cells to initiate DNA synthesis.

Self-Testing Questions

10. MPF activity controls the ability of cells to enter DNA synthesis.

11. Maturation-promoting factor and mitosis-promoting factor have the same function.

12. MPF is unique to frog eggs.

13. Lamins A, B, and C are proteins that are associated with the actin filaments that pull the chromosomes to different poles.

14. Active MPF has tyrosine-15 and threonine-161 phosphorylated.

15. Daughter cells of the yeast *S. cerevisiae* can be distinguished from their mother because the daughter cells are smaller.

16. Newly budded daughter yeast cells are in G_1.

17. The substrate specificity of cyclin-dependent protein kinases (**Cdk's**) is dictated by the specificity of the cation bound to the cyclin.

18. Checkpoint controls can, among other things, check for damaged DNA and repair it prior to DNA synthesis.

19. Inhibitors of actin polymerization, such as colchicine, are often used to prepare cells for karyotyping.

20. p53 **tumor suppressor** gene protein is linked to the G_1 checkpoint control.

Topic 12 *The Cell Cycle and DNA Repair*

Fill in the blanks of the following sentences to make accurate statements.

21. Chromosomes line up in a "plate" during _____.

22. Frog ovarian cells become arrested in the _____ phase prior to stimulation by progesterone.

23. Egg cells are also called _____.

24. Cyclin B degrades because of the presence of the destruction _____, a sequence of nine amino acids in the N-terminal end of the protein.

25. A decrease in _____ activity is necessary to exit mitosis.

26. The phosphorylation of _____ regulates the dissolution of the nuclear envelope.

27. A _____ is a small nuclear-envelope encapsulated chromosome seen prior to fusion with similar structures to form a mature nucleus.

28. Lamin repolymerization in the nuclear lamina is associated with _____ of lamin.

29. The cleavage furrow of cytokinesis occurs because of the **dephosphorylation** of _____.

30. Yeast **Cdc2**-**Cdc13** heterodimers can also be regarded as the *Xenopus* _____.

31. Cdc13 is also regarded as a _____.

32. _____ triggers the exit of yeast cells from G_1 into S.

33. _____ is a control point in yeast in G_1 after which yeast are committed to finish the rest of the cell cycle.

34. _____ can cause extra chromosomes to enter daughter cells. This can result in Down syndrome.

Select the letter that best completes the following statements.

35. The last phase of mitosis, during which two daughter cells are formed, is called _____.
   (a) telophase
   (b) anaphase
   (c) metaphase
   (d) prophase
   (e) pro-metaphase

36. Which of the following cell cycle phases is the shortest in duration? _____
   (a) G_1
   (b) G_0
   (c) G_2
   (d) S
   (e) M

37. What is the approximate time for one complete cell cycle of a typical mammalian cell? _____
   (a) 1 minute
   (b) 10 minutes
   (c) 5 hours
   (d) 24 hours
   (e) 5 days

38. Which of the following was discovered in egg cytoplasm and leads to the maturation of egg cells *in vitro* when injected into the cytoplasm? _____

(a) mitosin
(b) maturation-promoting factor
(c) Cdk inhibitor
(d) lamin
(e) kinetochore

39. MPF _____.

(a) is a protein kinase
(b) is a DNA repair enzyme
(c) is a protein that maintains the same level during the cell cycle
(d) is controlled by DNA polymerase I
(e) all the above

40. MPF is a heterodimer that includes _____.

(a) cyclin
(b) cAMP-dependent protein kinase
(c) Ced proteins
(d) histone H1
(e) all the above

41. Cyclin is synthesized continuously through the cell cycle but is quickly destroyed during _____.

(a) G_2
(b) anaphase
(c) G_1
(d) G_0
(e) none of the above, i.e., cyclin is not destroyed during the cell cycle.

42. Cyclin B is marked for destruction through the attachment of _____.
   (a) actin
   (b) cadherin
   (c) tubulin
   (d) clathrin
   (e) ubiquitin

43. MPF regulates all the following except _____.
   (a) condensation of chromosomes
   (b) glycolysis
   (c) lamin phosphorylation
   (d) dissolution of the nuclear envelope
   (e) formation of the mitotic spindle

44. Nuclear lamins are members of which family? _____
   (a) actin family
   (b) collagen family
   (c) intermediate family
   (d) microtubular family
   (e) not known

45. Yeast *cdc* mutants are easily recognized in the scanning electron microscope because _____.
   (a) their cells are smaller then wild-type
   (b) their cells are much longer than wild-type
   (c) they have microvilli present on the outside
   (d) they have multiple cleavage furrows on their surface
   (e) all the above

46. Cdc2 is _____.
   (a) a disaccharide
   (b) a protein kinase
   (c) a cell death gene
   (d) present only during interphase in yeast cells
   (e) none of the above

47. The restriction point in mammalian cells is similar to which of the following in yeast cells? _____
   (a) S block
   (b) ligation point
   (c) START
   (d) checkpoint alpha
   (e) all the above

48. Cdk1 is the same as _____.
   (a) **Cdc1**
   (b) Cdc2
   (c) Cdc3
   (d) Cdc4
   (e) Cdc5

49. E2F is _____.
   (a) a mitotic tubulin
   (b) attached to the kinetochore
   (c) a transcription factor that stimulates enzyme synthesis for S phase
   (d) an inhibitor of cyclin E
   (e) an inhibitor of Cdk4

50. Many of the studies done on the cell cycle were accomplished using egg cells from marine animals and amphibians. Why are these suitable organisms for this type of study?

51. Other studies done on the cell cycle have been accomplished using yeast such as *S. cerevisiae*. Why have they been suitable for cell cycle studies?

52. Cell fusion experiments have been extremely useful in identifying factors that regulate cell division. On the basis of the following cell fusion experiments, what nuclear events would you expect to see and how might you monitor this activity?

Fusion Cells	Effect	Manner of Investigation	Implication
(a) G$_1$ with S	_____	_____	_____
(b) G$_2$ with S	_____	_____	_____
(c) M with G$_1$	_____	_____	_____
(d) M with G$_2$	_____	_____	_____

53. You are asked to investigate the cell cycle of human fibroblasts. Your charge is to approximate the cell cycle compartment times for each compartment. To your initial dismay, you realize that the only probe that you have to facilitate this quest is ^{3}**H-thymidine**. How might you use only this probe to determine the cell cycle compartment times?

54. You are given an antibody to cyclin B and asked to monitor cyclin B activity in a group of synchronously dividing human keratinocytes (skin cells). You decide to use Western blots of native gels as your major tool for investigation. You get the following results:

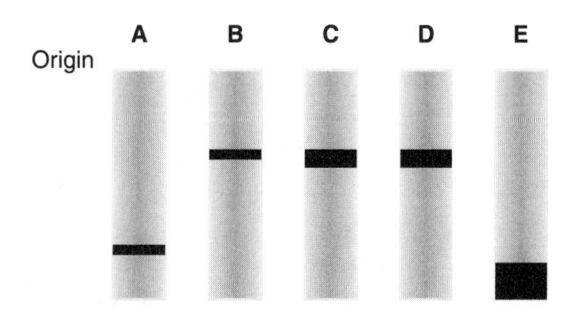

A – Purified cyclin
B – Cytoplasm from G2 cell
C – Cytoplasm from prophase cell
D – Cytoplasm from metaphase cell
E – Cytoplasm from late anaphase cell

(a) How can these results be interpreted?
(b) How could you determine whether cyclin is actually controlling mitosis through its cyclical appearance or is merely a consequence thereof?

(c) How might you assay for MPF activity during these cycle compartments?
(d) How could you determine if this apparent cycling of MPF activity is governed by nuclear or cytoplasmic events?
(e) How might you show that the rise in MPF activity during the cell cycle is due, in part, to the increased synthesis of cyclin as suggested by the Western blots?
(f) How might these Western blots differ if cyclin B lacks the "destruction box"?
(g) How might these Western blots differ if ubiquitin were unable to attach to cyclin B?
(h) Why might MPF activity stay high during mitosis when cells are treated with the microtubular inhibitor colchicine?

55. The MPF-induced phosphorylation of lamins appears to be critical in the dissolution of the nuclear envelope.
(a) How could you show that MPF phosphorylates lamins directly?
(b) How could you show that the phosphorylation of lamins, such as lamin A, is a critical step in the separation of daughter nuclei in anaphase?

56. Cyclin D and Cdk2 are critical for cells to go through the G_1 phase of the cell cycle. Draw a typical Northern blot and activity blot below, indicating the relationship between cyclin D mRNA and cyclin D-Cdk2 activity.

Northern blots of cyclin D mRNA
G_0 Early G_1 Middle G_1 Late G_1 Early S

Relative levels of cyclin D-Cdk2 protein kinase activity
G_0 Early G_1 Middle G_1 Late G_1 Early S

ATP

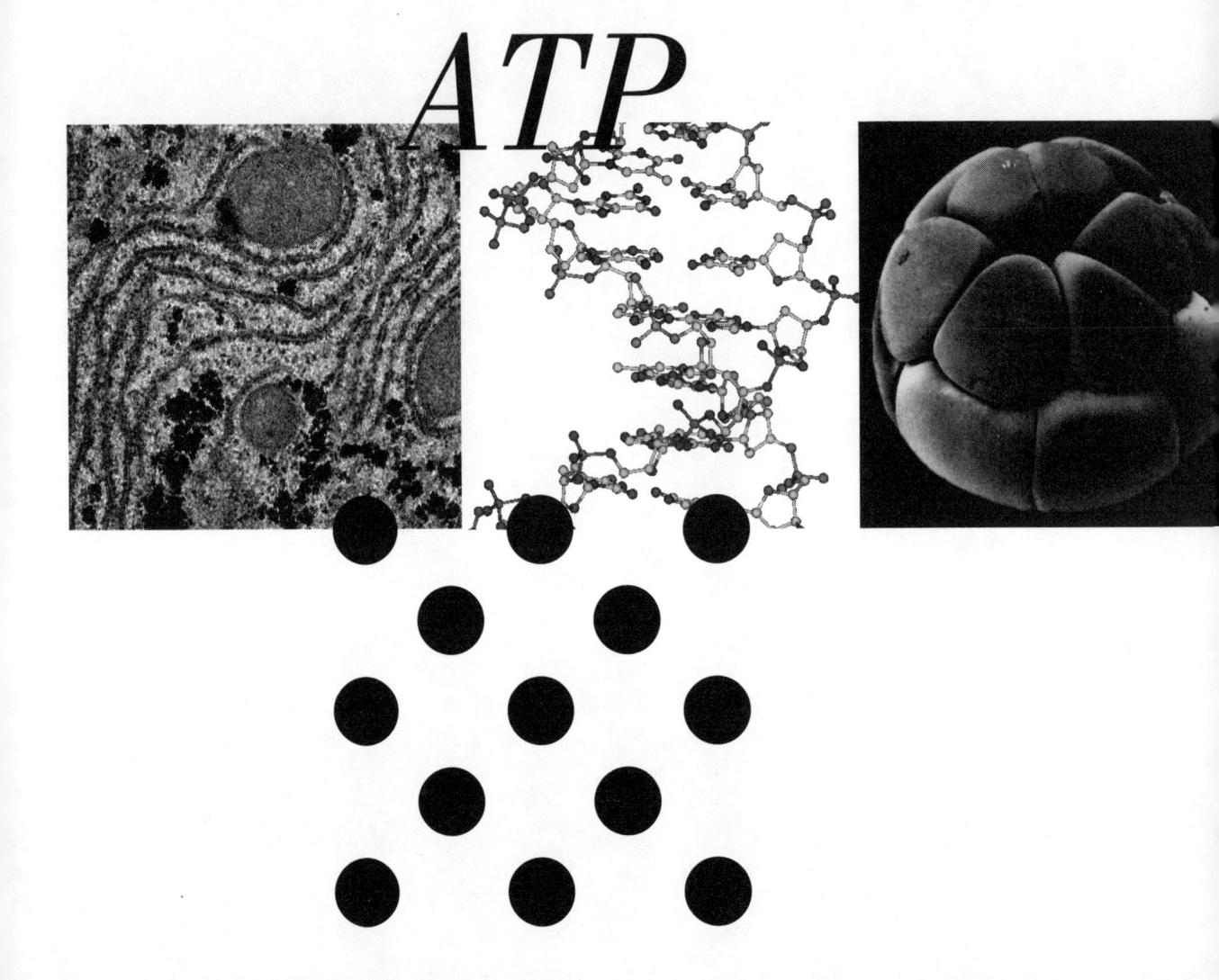

topic 13
Cancer and Cell Death

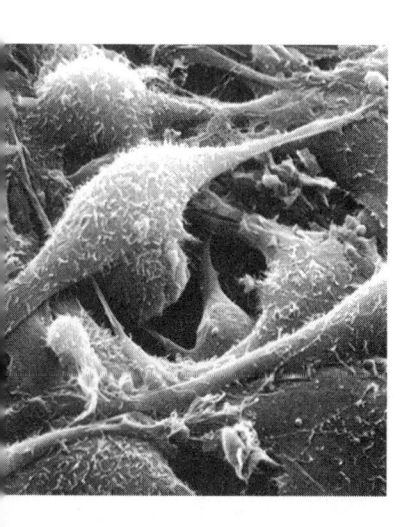

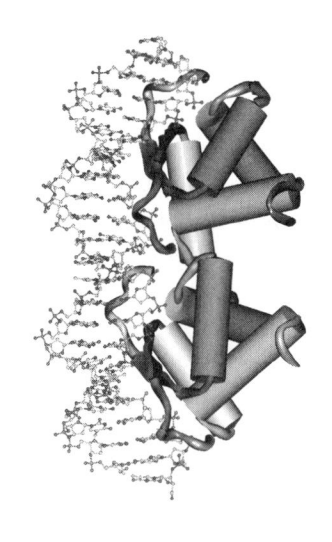

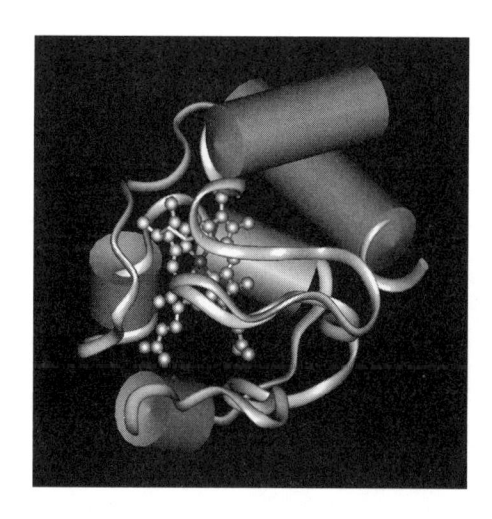

Summary

One of the biggest paradoxes in medicine is cancer. Cancer results from the unregulated control of cell division, and cancer cells are therefore very successful in their ability to propagate. Yet their success is often the demise of the organism that they inhabit. This topic explores the differences between cancer cells and normal cells, as well as how cells die through either **necrosis** or programmed cell death.

Cancer cells can be recognized from normal cells in many different ways. The best distinguishing difference between the two types is their growth in soft agar. Soft agar is a semi-solid medium that encourages the division of cancer cells but not most normal cells. Cancer cells survive better in soft agar because they are less dependent than normal cells on attaching to a substratum. Normal cells secrete fewer **proteases** than cancer cells and have a more organized cytoskeleton. Normal cells grown in culture are mortal, usually succumbing after 25 to 50 divisions. Cancer cells, by definition, are immortal and can divide continuously in culture as long as they are given a continuous supply of nutrients. Finally, cancer cells, but not normal cells, can cause tumors when injected into the appropriate animal host.

Normal cells can be converted to cancer cells through a variety of mechanisms. Radiation such as ultraviolet light and X rays can cause mutations in DNA that in turn can cause unregulated cell division. Viruses can cause a variety of human cancers by introducing their DNA into the human genome and altering its function. Chemical **carcinogens** in the environment can cause cancer by binding to bases. A few compounds that are not carcinogens can be mistakenly converted to carcinogens in the liver. Spontaneous mutations can occur that result in a base change. If this alteration is not corrected by one of the myriad DNA correction enzymes, it could lead to altered genetic activity and cancer. No matter how a normal cell is converted to a cancer cell, however, some stable type of gene alteration must occur. One of these changes is the conversion of a **proto-oncogene**, a gene in normal cells that functions during cell growth and division, to an active **oncogene**, a gene whose activity results in cell **transformation** from the normal phenotype to a cancer cell.

A variety of oncogenes that have been identified. Some are growth factor oncogenes such as the *sis* oncogene, whose product is platelet-derived growth factor. Other oncogenes have protein products that are in the form of altered growth factor receptors. An example is the *erbB* oncogene, whose oncoprotein is a modified version of the epidermal growth factor. Many of these **oncoproteins** are receptors that can't be regulated by an external stimulus and, as such, are constitutively turned on. Other oncogenes work as intracellular transducers. One such oncogene, *src*, produces an oncoprotein that, as a **protein kinase**, phosphorylates tyrosine residues. Other oncogenes in this family work through *Ras*. Nuclear transcription factor oncogenes also exist by regulating transcription. But the group of oncogenes that causes more than half of all human cancers belong in the cell cycle control group and are called tumor suppressor genes. The tumor suppressor gene p53 works as a checkpoint in the G_1 part of the cell cycle. If damaged DNA is detected, it stops the cycle and repairs it before the cycle can proceed. People with defective p53 are unable to repair the DNA, and if this damaged DNA is replicated, cells could become transformed to cancer cells.

The opposite of cancer could be considered cell death. Cell death can occur in two fundamentally different manners. Cells can die through a pathological event, a process called *necrosis*. Necrotic cells are characterized by an abrupt increase in size and leaky membranes. No gene activation and no special input of ATP are necessary for cells to die in this manner. Apoptosis, or programmed cell death, is different from necrosis. Apoptosis occurs through activation of cell death genes, many of which have been identified in both vertebrates and invertebrates. Apoptosis causes the cell to shrink, with many plasma membrane blebs emerging from the cell. The nucleus is fragmented and the **chromatin** condenses. Neighboring cells often endocytose the apoptotic cell. The DNA is cleaved in a nonrandom manner, revealing a DNA ladder on a gel.

While cancer and cell death appear to be opposites of each other, we are now recognizing that many chemotherapeutic treatments, such as radiation, that have been known for decades to kill cancer cells do so through activation of apoptosis. Thus, there is new hope and new energy devoted to understanding how apoptosis might be selectively activated in cancer cells—a strategy that is best accomplished once the biochemical basis of both cancer and apoptosis are fully appreciated.

Self-Testing Questions

Fill in the blanks of the following sentences to make accurate statements.

1. _____ cells often originate from a malignant tumor and travel through the bloodstream or lymphatic system, where they cause another tumor.

2. The most common type of human cancer is _____.

3. _____ is a cancer of the pigmented skin cells and much more deadly than its relative, the basal-cell **carcinoma**.

4. Phorbol esters can activate **protein kinase C** and are considered tumor _____ because they can cause cancer only after a different chemical has been used as the tumor initiator.

5. The _____ viruses can cause warts.

6. Cancer cells can cross the basal lamina and invade other tissue areas. They do this by secreting the enzyme _____.

7. The _____ gene codes for an ATP-hydrolyzing plasma membrane transporter that pumps lipophilic cancer drugs out of cells, making cancer cells more resistant to chemotherapeutic agents.

Self-Testing Questions

8. A proto-oncogene can be converted to a cancer-causing _____.

9. A defective _____ gene such as *p53* or *RB* can cause cancer.

10. The _____ oncogene was isolated from the Rous sarcoma virus.

11. An RNA virus is also called a _____.

12. Normal cells that incorporate an oncogenic virus and are converted to cancer cells are considered to have been _____.

13. Membrane blebs are characteristic of cells going through programmed cell death, otherwise known as _____.

14. *bcl-2* inhibits _____.

15. DNA agarose gels can be used to detect _____ cells because the DNA banding pattern takes on a ladder-like appearance.

Select the letter that best completes the following statements.

16. Cancer can be classified in many different ways. Which of the following is a cancer arising from epithelial cells? _____
    (a) leukeumia
    (b) sarcoma
    (c) carcinoma

17. Cancers can be caused by DNA mutation or a change in the pattern of gene expression. The latter types of cancers are referred to as _____.
    (a) epigenetic
    (b) morphologic
    (c) phenotypic
    (d) abnormal
    (e) none of the above

18. Which of the following can convert noncancer cells to cancer cells?_____
    (a) irradiation
    (b) chemical carcinogens
    (c) oncogenes
    (d) defective tumor suppressor genes
    (e) all the above

19. Aflatoxin is an indirect-acting carcinogen because it can be converted to an active carcinogen, aflatoxin-2,3-epoxide, by which type of enzyme system? _____
    (a) plastocyanin
    (b) flavin mononucleotide
    (c) cytochrome P_{450}
    (d) p700
    (e) NADH

20. Liver extracts are often used with bacterial cultures to detect carcinogens because the liver extract can _____.
    (a) nourish the bacteria so they grow better
    (b) modify indirect carcinogens
    (c) be used as a marker for mutant bacteria
    (d) both parts (a) and (b)
    (e) none of the above

Self-Testing Questions

21. The bacteria test used to detect chemical carcinogens was developed by _____.
- (a) Peter Mitchell
- (b) Robert Emerson
- (c) Bruce Ames
- (d) Daniel Branton
- (e) John Dowling

22. Xeroderma Pigmentosum, HNPCC syndrome, and ataxia telangiectasia are all cancers caused by _____.
- (a) delayed G_1 phase of the cell cycle
- (b) kinetochore abnormalities
- (c) centromere abnormalities
- (d) DNA repair enzyme defects
- (e) unknown causes

23. Which of the following can be an *in vitro* characteristic of a cancer cell, but not of a normal cell?_____
- (a) large dependence on serum
- (b) high cytoplasmic to nuclear ratio
- (c) secretion of high levels of extracellular proteases
- (d) secretion of high levels of extracellular matrix components
- (e) flat phenotype

24. One way to subclone cancer (but not normal) cells *in vitro* is _____.
- (a) growth in soft agar
- (b) growth on Type IV collagen-coated culture dishes
- (c) growth on Type I collagen-coated culture dishes
- (d) growth on laminin-coated culture dishes
- (e) growth on hyaluronic acid

25. Which of the following is often characteristic of cancer cells but not of normal cells?____
  (a) enhanced transport of metabolites
  (b) increased plasma membrane mobility
  (c) poor organization of actin filaments
  (d) immortality
  (e) all the above

26. Viral oncogenes such as v-*src* probably originated from ____.
  (a) cellular proto-oncogenes
  (b) other viruses that fused with the virus containing v-*src*
  (c) bacteria
  (d) mutation of viral genes
  (e) none of the above

27. Proto-oncogenes can be converted into active oncogenes by ____.
  (a) deletion mutation
  (b) point mutation
  (c) gene amplification
  (d) chromosome rearrangement
  (e) all the above

28. *src* is most similar to ____.
  (a) tyrosine protein kinases
  (b) nuclear transcription factors
  (c) GTP-binding proteins
  (d) growth factors
  (e) growth factor receptors

Indicate whether the following statements are True or False.

29. Benign tumors are composed of cells that are often neoplastic but remain clustered together and are not **metastatic**.

30. Adenomas are generally of less clinical concern than adenocarcinomas because the former are benign tumors, whereas the latter are malignant tumors.

31. Mutagenesis is a change in DNA that can lead to cancer.

32. Tumor cells typically are mosaics, with some cells showing one X chromosome inactivated and other cells showing the other X chromosome inactivated.

33. The chance of contracting cancer decreases with age.

34. A tumor initiator often predisposes DNA for oncogenesis once the cell has been exposed to a tumor promoter.

35. Oncogenes can cause cancer only if both copies of the normal-allele proto-oncogene are converted to oncogenes.

36. Tumor suppressor genes can cause cancer only if both tumor suppressor gene alleles are inactivated.

37. Retinoblastoma occurs only if a defective *RB* gene is inherited.

38. The *RB* gene is a tumor suppressor gene.

39. People who inherit one defective p53 gene are predisposed to cancer.

40. p53 works by acting as a brake on cell division if damaged DNA is detected.

Topic 13 Cancer and Cell Death

41. Apoptotic cells are characterized by leaky membranes.

42. An underexpression of p53 can cause programmed cell death.

43. Necrosis can be caused by activation of the *ced-3* and *ced-4* genes in *C. elegans*.

44. Growth curves of cells in culture can be diagnostic of oncogenic activities. Shown below are graphs of various types of cells growing in standard culture plates. Explain what the curves might suggest about the different cell cultures. Assume in all cases that the initial seeding densities are the same and all culture plates are identical unless otherwise specified. Also assume that the cell cycle time in all cases is 24 hours.

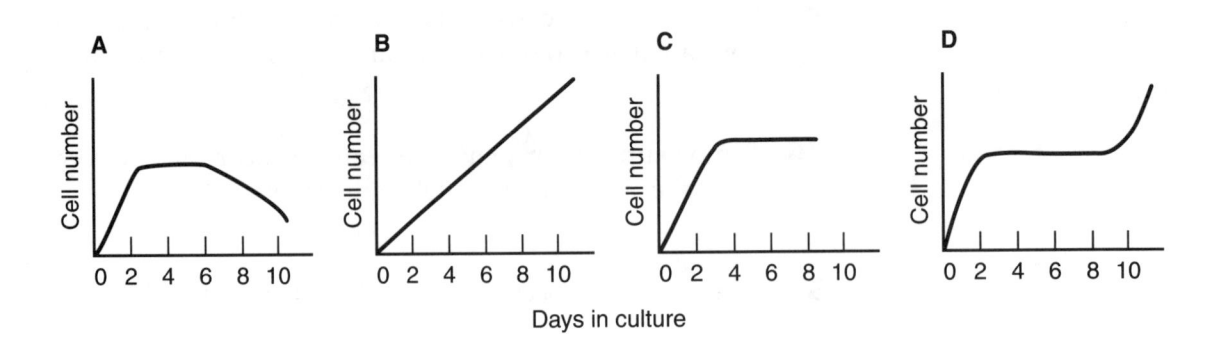

45. Growth curves are often useful for examining the carcinogenic potential of chemicals. You set up a special experiment that consists of three cell culture conditions. In graph A, normal 3T3 cells are grown in culture and reach confluence at day 4. Graph B is the same condition as A but with a putative carcinogen added. Graph C is the same as B but with a co-culture of contact inhibited human liver (hepatocyte) cells grown on a microporous membrane support. How might the growth kinetics be explained?

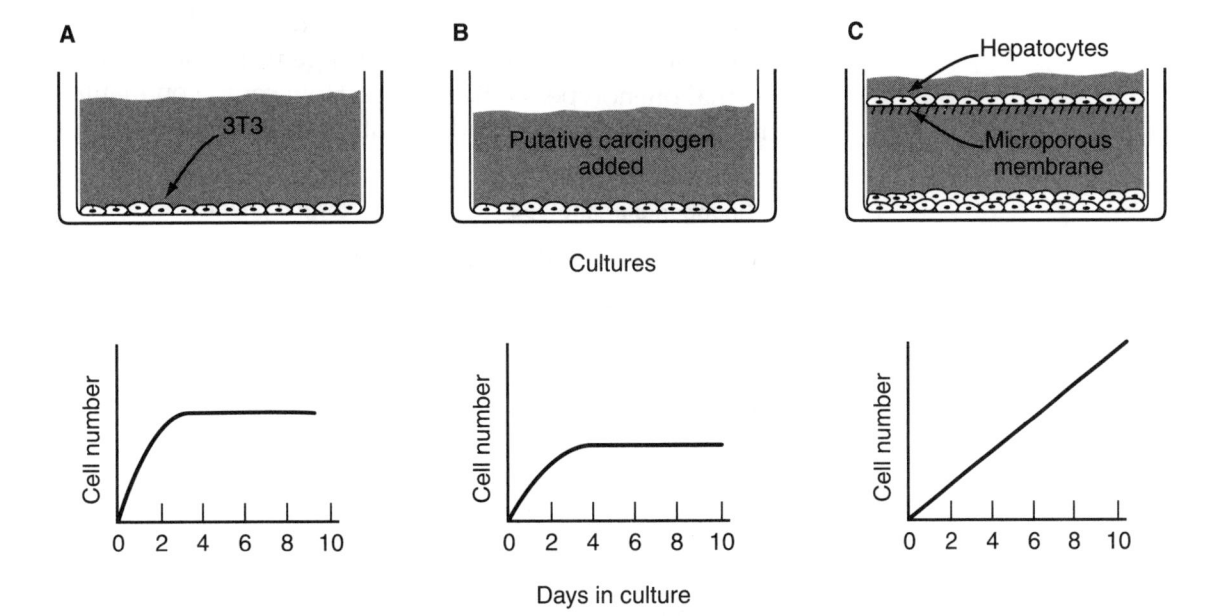

46. Growth curves are also useful for analyzing cell death characteristics. As shown in graph A, sympathetic neurons can grow, but usually don't divide, in culture in the presence of nerve growth factor (NGF). As shown in B, withdrawal of NGF causes the cells to die. Yet, as shown in C, NGF withdrawal with the simultaneous addition of a protein synthesis inhibitor results in different growth kinetics. How might the observation in C be explained, and what is the implication for necrosis and apoptosis?

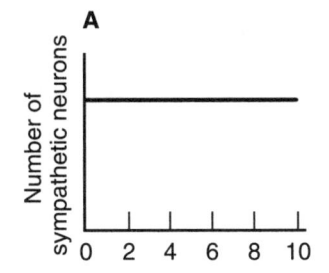

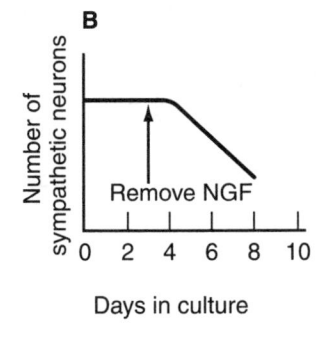

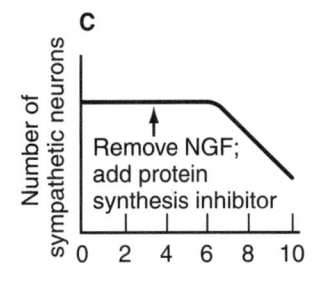

Topic 13 *Cancer and Cell Death*

47. Some cells can be grown under different conditions that allow them to express either an oncogenic phenotype or a normal phenotype, as shown here. How could you manipulate the cells to express these different conditions?

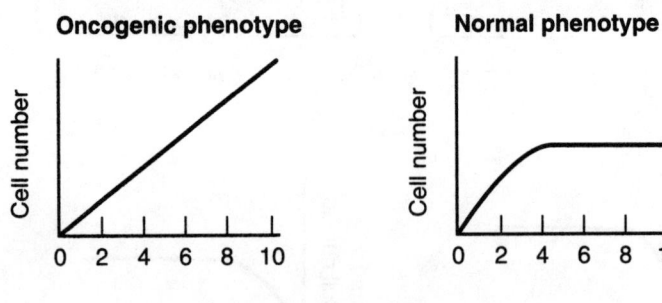

Oncogenic phenotype

Normal phenotype

Days in culture

48. Experiments have been reported in the literature showing the apparent peculiar activity of fused cells in culture. It is possible to fuse a normal mouse cell with a human cancer cell, resulting in a hybridoma that initially has a normal phenotype in culture (see graph A below). Yet after propagation of these cells for many generations, the hybridoma takes on the oncogenic phenotype. How might this phenomenon be related to tumor suppressor genes?

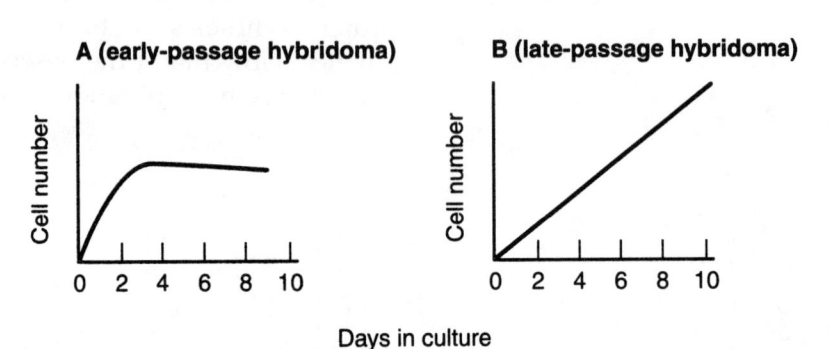

A (early-passage hybridoma)

B (late-passage hybridoma)

Days in culture

49. The *src* oncogene is transfected into a normal fibroblast and causes the cells to take on the oncogenic phenotype. A site-directed mutant of the *src* oncogene is constructed whose oncoprotein lacks the ability to have myristic acid added to it post-translationally. **Transfection** of this mutant into the normal fibroblasts results in growth that is characteristic of normal cells, not cancer cells. How might this be explained?

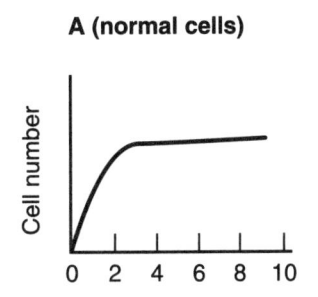

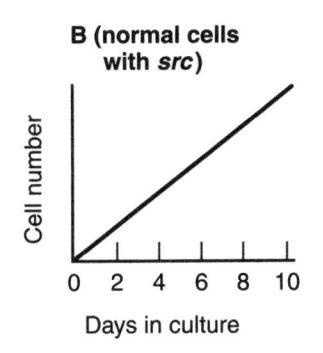

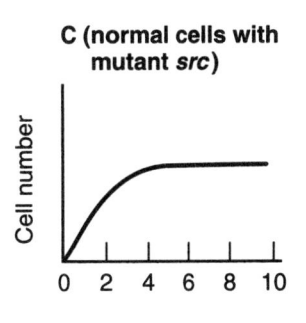

A (normal cells)

B (normal cells with *src*)

C (normal cells with mutant *src*)

Days in culture

50. Phorbol esters are nasty chemicals that can cause cancer. If we treat some cells with a phorbol ester, we find that they temporarily take on a cancer phenotype as shown in growth curve A. Soon, though, they revert to normal cells. If, on the other hand, the cells are pretreated by a putative carcinogen that does not, by itself, cause the cells to become cancerous, the phorbol esters now appear to produce stable cancer cells (curve B). How can the differences in these growth curves be explained?

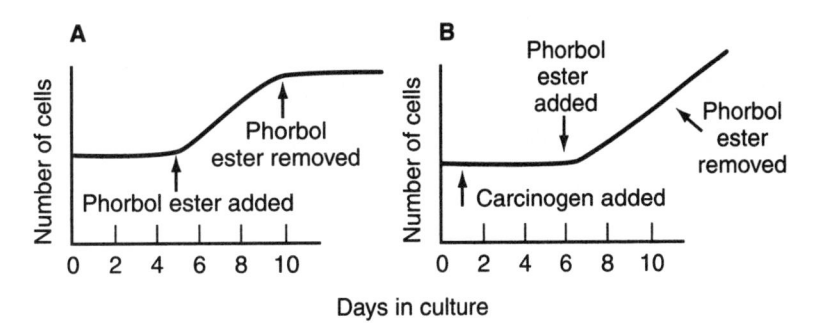

A — Phorbol ester added, Phorbol ester removed

B — Carcinogen added, Phorbol ester added, Phorbol ester removed

Days in culture

Topic 13 Cancer and Cell Death

51. Isolate a possible cancer cell type from a tumor and grow it in culture under two different conditions. Condition A is in standard cell culture medium, whereas condition B is in soft agar. What do these data suggest about the oncogenic nature of this new cell type?

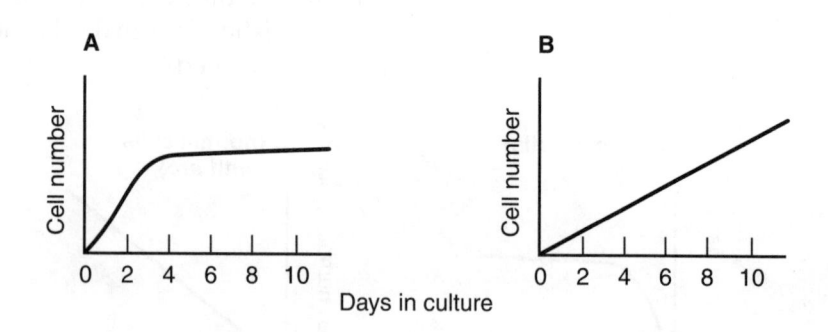

A

Cell number vs. Days in culture (0 2 4 6 8 10): rapid rise then plateau.

B

Cell number vs. Days in culture (0 2 4 6 8 10): steady linear increase.

52. Cultured skin cells that are stored at 4°C for a few days and then brought to 37°C undergo a strange behavior in their metabolic activity. While they appear metabolically fine when first brought to 37°C, they die over the course of a week, as shown in the figure below. How might you determine whether this phenomenon is due to necrosis or apoptosis?

Relative metabolic activity (i.e., cell viability) vs. Days of recovery at 37°C (0 2 4 6 8 10 12): gradual decline.

ATP

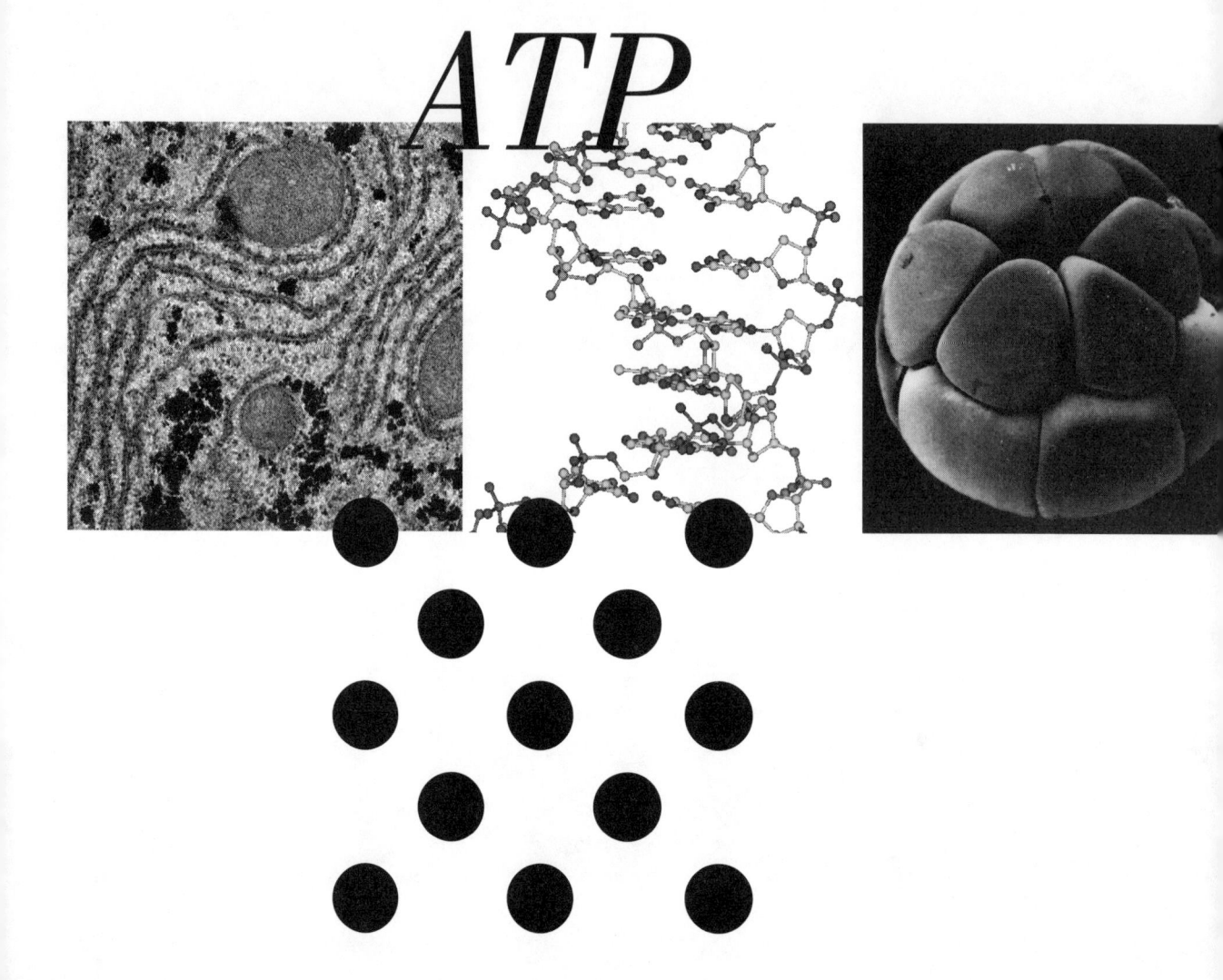

topic 14
Development and Differentiation

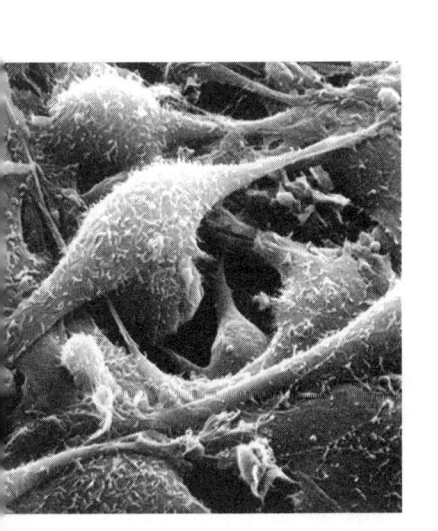

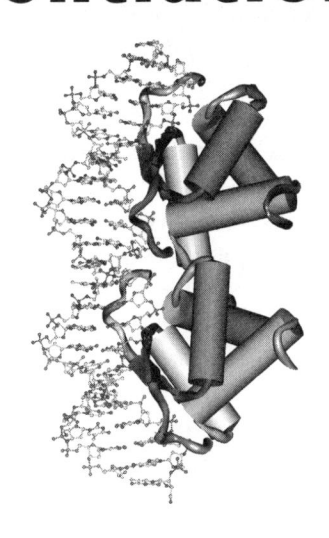

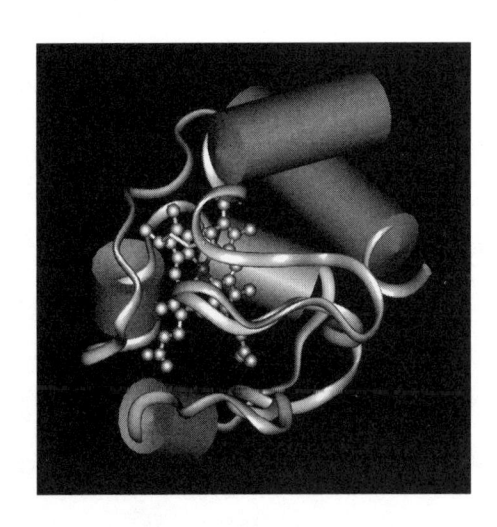

Summary

Development and cell differentiation are two of the most fascinating processes in nature. Development is defined as the progressive change of an organism as it moves from a fertilized egg to the mature adult. Differentiation, on the other hand, is the process of change that a stem or progenitor cell undergoes as it becomes a more specialized cell. Both development and differentiation are governed by coordinated gene activities. While many organisms have been used to study development, the amphibian *Xenopus laevis*, the fruit fly *Drosophila*, and the roundworm *C. elegans* have been most actively studied.

The development of the amphibian embryo consists of two major phases. In the first phase the fertilized egg develops into a **blastula**, then a **gastrula**, and finally a neural stage. In the second phase, organogenesis (the development of organs) occurs. The amphibian egg has a polarity that is distinguished by the animal and vegetal poles. The egg is fertilized, generating a **gray crescent** opposite the point of sperm penetration. The fertilized egg divides into **blastomeres**, the constituent cells of the blastula. At the 16-cell stage, water flows into the blastula, creating a lumen (blastocoel) within this multicelled sphere. During the gastrula phase the basic body plan is constructed. Cells turn inward to form the dorsal lip of the **blastopore**. This inward movement results in a future gut with three layers of cells: the **ectoderm**, **mesoderm**, and endoderm. The **Spemann's organizer** is found in the dorsal lip. This patch of cells secretes an embryonic signal that directs the differentiation of cells. An extra organizer transplanted into an embryo can cause a double-headed animal. The **notochord** develops, which is the precursor of the vertebral column. The neural stage starts at the end of **gastrulation** and is characterized by ectoderm that rolls up into a tube on the dorsal side of the embryo. The formation of this neural tube, which is the future brain and spinal cord, requires actin filaments. Much of this **morphogenesis** is induced by **morphogens**—chemicals that are released from a single site and diffuse in a radial pattern. They induce differentiation and inform the adjacent cells of their position within the embryo.

Another well-studied animal is *C. elegans*. It is a nonparasitic roundworm whose excellence as a developmental model was first emphasized by Robert Horvitz of the Massachusetts

Institute of Technology. *C. elegans* is translucent, easily grown, and has a short generation time. It is easily microinjected, and composed of a limited number of cells. Most importantly, however, the lineage of each cell has been described in detail. Studies in *C. elegans* have shown that in some cases a single "founder" cell can give rise to all cells within a particular group. This is the case, for instance, with intestinal cells. Muscle cells, however, arise from different cell clones. Experiments using laser ablation have shown that the anchor cell induces hypodermal cells to become a vulva. Through the study of mutant *C. elegans*, several genes have been identified whose dysfunction results in no vulva. Different studies with *C. elegans* have identified cell death genes and cell death inhibitor genes. The cell death genes *ced-3* and *ced-4* are important during programmed cell death of the 131 cells that succumb as a natural process of development. *ced-9* is a cell death inhibiting gene and is homologous to the mammalian cell death inhibitor gene *bcl-2*.

Drosophila has been used for decades as a model organism to study development. It goes through a stereotypical pattern of developmental stages that include **fertilization**, hatching, larval, pupation, and the adult. Of particular interest in *Drosophila*, but not the other two model organisms mentioned previously, are polytene chromosomes. These are chromosomes with highly duplicated DNA and a distinct banding pattern. Mutations can be detected on polytene chromosomes through simple microscopy. This permits the approximate localization of the defective gene and, so, aids in its molecular characterization.

In summary, studies of *Xenopus laevis*, *C. elegans*, *Drosophila*, and other organisms have yielded insight into general principles governing development and differentiation.

Self-Testing Questions

Fill in the blanks of the following sentences to make accurate statements.

1. The large volume of the *Xenopus* egg is occupied primarily by _____ platelets.

2. The yolk in the *Xenopus* egg is concentrated in the _____ pole, whereas the _____ pole is the top side of the egg.

3. After the sperm penetrates the *Xenopus* egg, some of the pigmentation of the *Xenopus* egg directly opposite the point of sperm entry moves, creating the _____.

4. The _____ are cells that form immediately after fertilization of the *Xenopus* egg. They are characterized by rapid cell division, but with no concomitant increase in size.

5. A _____ map shows the body part each cell is destined for from the gastrula stage of amphibian embryogenesis on.

6. The _____ organizer is located at the dorsal lip of the _____ of the amphibian embryo.

7. The _____ is a specialization of the mesoderm and is the precursor of the vertebral column.

8. The _____ tube forms the brain and the spinal cord in the amphibian embryo.

9. Muscle cells in developing vertebrates come from the _____.

10. _____ is the process by which one group of cells influences the direction of differentiation of neighboring cells.

Self-Testing Questions

11. Vg1 proteins in the *Xenopus* embryo induce the _____ at the dorsal lip of the blastopore.

12. The _____ surrounds the early developing mouse embryo.

13. A _____ is an individual that forms from two genetically distinct groups of cells.

14. Single cells isolated from the eight-stage mammalian embryo can generate eight totally normal individuals. Thus these cells are regarded as _____.

15. _____ analysis of *C. elegans* has revealed the origin and fate of each cell of the organism.

16. The intestine of *C. elegans* originates from a single _____ cell.

17. The _____ cell regulates the development of the *C. elegans* vulva.

18. A _____ mutation in *C. elegans* causes cells to behave in an "age-inappropriate" manner (i.e., as if they were earlier in the cell lineage).

19. Programmed cell death or suicide is also called _____.

20. Fertilized *Drosophila* eggs develop many nuclei in a common cytoplasmic mass called a _____.

Topic 14 *Development and Differentiation*

21. The egg polarity genes in *Drosophila* help regulate _____ , chemicals that, once released from one site, can induce the differentiation of cells at a different site.

22. Polar granules are segregated and taken up by some *Drosophila* and *C. elegans* cells so that cells containing these granules will develop into _____ line cells.

Select the letter that best completes the following statements.

23. The vegetal pole of the *Xenopus* egg will primarily form the _____.
    (a) skin
    (b) liver
    (c) pancreas
    (d) reproductive organs
    (e) spleen

24. The back of the animal is often referred to as _____.
    (a) ventral
    (b) anterior
    (c) posterior
    (d) dorsal
    (e) medial

25. The front of the developing embryo is often referred to as _____.
    (a) ventral
    (b) anterior
    (c) posterior
    (d) dorsal
    (e) medial

26. Immediately after fertilization of the *Xenopus* egg there are a number of movements in the egg that are mediated by _____.
   (a) intermediate filaments
   (b) actin filaments
   (c) microtubules
   (d) lamins
   (e) all the above

27. Blastomeres in the *Xenopus* blastula are coupled together by tiny channels that can pass small molecules. These membrane specializations are called _____.
   (a) tight junctions
   (b) desmosomes
   (c) cadherins
   (d) gap junctions
   (e) none of the above

28. The initial formation of the amphibian gut is most closely associated with the _____.
   (a) gastrula
   (b) blastula
   (c) neural tube
   (d) gray crescent
   (e) none of the above

29. Mesenchyme in the developing amphibian embryo is found in the _____.
   (a) epithelium
   (b) ectoderm
   (c) mesoderm
   (d) endoderm
   (e) none of the above

30. The formation of the blastopore in the *Xenopus* embryo is most often associated with which of the following stages? _____

(a) fertilized egg
(b) neural tube
(c) blastula
(d) gastrula
(e) none of the above

31. Musculature originates primarily from _____.

(a) endoderm
(b) mesoderm
(c) ectoderm
(d) not known

32. Which of the following are signaling molecules involved in the induction of the mesoderm in *Xenopus laevis*? _____

(a) noggin
(b) FGF
(c) activin
(d) all the above
(e) none of the above

33. Cancers containing undifferentiated cells (i.e., stem cells) that can be transplanted from host to host are called _____.

(a) sarcomas
(b) carcinomas
(c) teratocarcinomas
(d) adenomas
(e) adenocarcinomas

34. Cells that have launched into a specific developmental program are called _____.

(a) stem cells
(b) dedifferentiated
(c) committed
(d) a teratoma
(e) determined

35. C. elegans is composed of how many cells? _____
   (a) 10
   (b) 100
   (c) 1000
   (d) 10,000
   (e) 100,000

36. A mutant *lin-45* C. elegans gene results in roundworms with no _____.
   (a) gut
   (b) anus
   (c) musculature
   (d) mouth
   (e) vulva

37. The effect of the C. elegans *lin-45* gene is to _____.
   (a) keep cells in an immortal state
   (b) initiate vulva development
   (c) initiate gut development
   (d) initiate **determination** in all cells
   (e) none of the above

38. Overactive *ced-3* C. elegans genes cause _____.
   (a) apoptosis
   (b) necrosis
   (c) differentiation
   (d) dedifferentiation
   (e) cell transformation (cancer)

39. Bcl-2 is _____.
   (a) a cell death gene discovered in C. elegans
   (b) a protein
   (c) a homologue of *ced-3*
   (d) a homologue of *ced-4*
   (e) none of the above

40. The generation time of *Drosophila* is approximately _____.
   (a) 1 day
   (b) 9 days
   (c) 45 days
   (d) 1 year
   (e) 2 years

41. The polar phenotype of the *Drosophila* embryo is determined by _____.
   (a) the mother
   (b) the father
   (c) both the mother and the father
   (d) factors not yet known

42. The *Drosophila* dorsal protein _____.
   (a) sets up a gradient in the embryo
   (b) in the dorsal blastoderm is present in the cytoplasm
   (c) in the ventral blastoderm is present in the nucleus
   (d) all the above
   (e) none of the above

43. The *bicoid* mRNA from *Drosophila* that codes for an anterior morphogen gradient is originally synthesized by _____.
   (a) helper cells
   (b) nurse cells
   (c) follicle cells
   (d) blastomeres
   (e) gastrula cells

44. The *bicoid* genes in *Drosophila* regulate _____.
   (a) the segmentation genes
   (b) the second-messenger genes
   (c) the pigmentation genes
   (d) the length of the antennae
   (e) none of the above

Indicate whether the following statements are True or False.

45. *Xenopus laevis* is an amphibian whose embryological development has been studied extensively.

46. The point of sperm entry into a *Xenopus* egg is roughly the future dorsal aspect of the animal.

47. Calcium is pumped into the interior of the blastula, water follows, and the blastocoel forms.

48. The gastrula stage of *Xenopus* follows the neural stage.

49. Invagination is a process by which cells turn inward in the embryo.

50. Convergent extension in *Xenopus* embryogenesis occurs in blastomeres.

51. The endoderm is the forerunner of a gut in the amphibian gastrula stage.

52. The *Xenopus* neural crest forms the central nervous system.

53. The vertebrate **retina** is derived from the neural tube.

54. The *Xenopus* egg nucleus is different from all other nuclei in the adult amphibian because the egg nucleus contains very different genes from the other nuclei.

55. Up to the 128-cell stage, any cell of the mammalian embryo can form a normal adult.

56. Mammalian development is highly regulative.

57. *C. elegans* are considered either hermaphrodites or males.

58. *C. elegans* has about the same number of genes as humans.

59. A "housekeeping" gene is one that regulates critical activities in all cells that are related to division and survival.

60. *lin-3* and *let-23* are two *C. elegans* genes that regulate the development of the anus.

61. An imago is another name for a larval, first instar *Drosophila*.

62. Treatment with gene transcription inhibitors will block cell division in most animal somatic cells *in vitro*, but cell division of blastomeres in the *Xenopus* eggs still continues in the presence of these inhibitors. Why is there such a difference between the two systems?

63. You are interested in the molecular explanation for how the blastocoel forms in the *Xenopus laevis* blastula. How could you determine if Ca^{2+}, Na^+, K^+, or Cl^- is involved in this process?

64. If amphibian embryos are allowed to develop in solutions containing EGTA or EDTA (calcium chelators), their development is often severely altered. What is the most probable reason underlying this alteration?

65. Integrins are integral plasma membrane proteins that are linked to the extracellular matrix and also linked to cytoskeletal elements inside of the cell. How might you show that integrin receptors and their association with fibronectin, an extracellular matrix component, are important to the development of the amphibian embryo?

66. How could you show that neural crest cells migrate from the dorsal ectoderm to other distant parts such as the adrenal gland?

67. Ultrasound, centrifugation, and other mechanical disturbances of the unfertilized *Xenopus* egg can dramatically affect the correct formation of the subsequent embryo. What is the basis for this alteration?

68. How can transplantation experiments demonstrate the importance of determination?

69. Determination is coupled with the expression of specific genes during development. How can *in situ* **hybridization** be useful in studying the time- and position-dependent expression of these genes?

70. Determination in the embryo can occur through a nuclear, cytoplasmic, or paracrine mechanism. How can transplantation experiments be used to confirm a suspected case of cytoplasmic memory or paracrine memory?

71. How might you use a laser to demonstrate that the anchor cell controls vulva development?

ATP

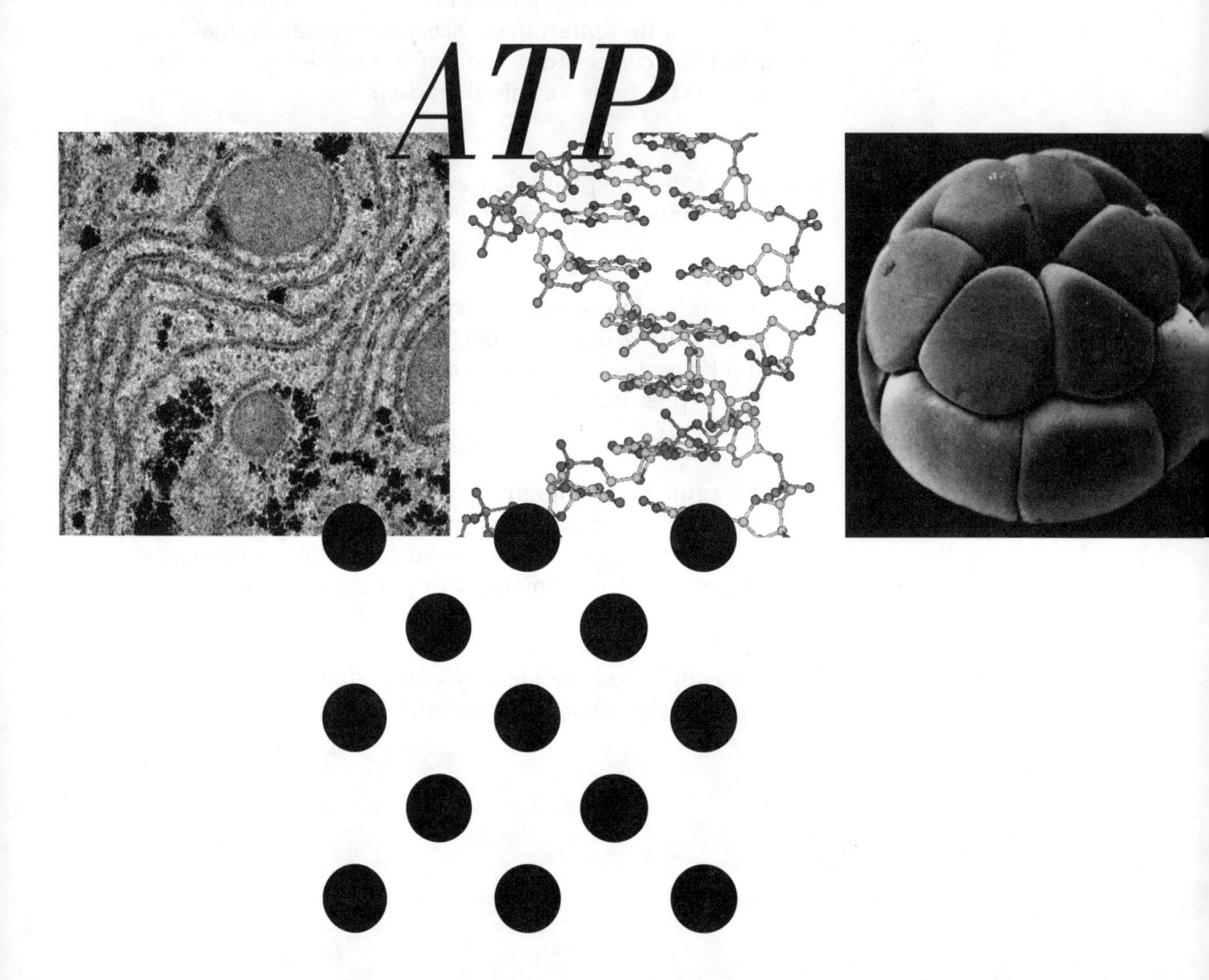

topic 15
Gene Expression in Eukaryotes & Prokaryotes

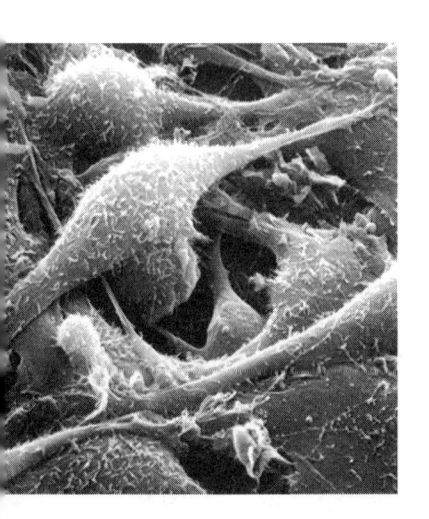

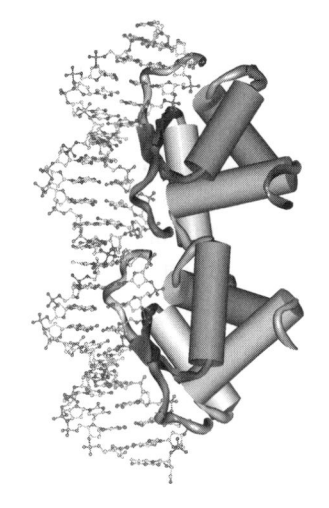

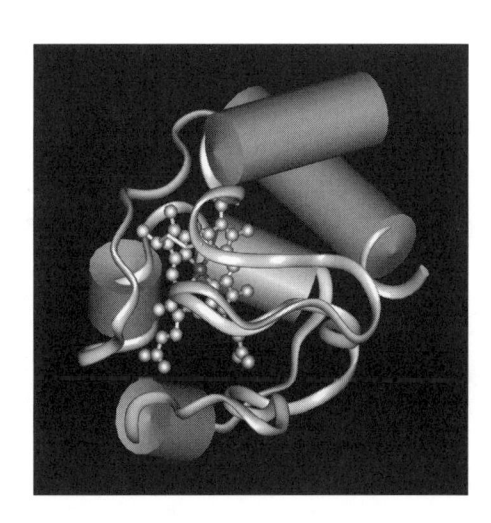

Summary

The morphological and functional properties of a cell depend largely on the proteins it contains. These proteins are encoded by the thousands of genes contained in the genome. Only a certain carefully selected number of these genes are expressed in any given cell type. What determines the amounts and types of protein expressed in each cell is the subject of this chapter.

While the amount of a particular protein in a cell can be regulated at many levels, transcription initiation is the primary form of regulation. Regulatory pressure may also be exerted at the elongation phase of transcription as well as at the translation step of mRNA.

In single-celled organisms such as bacteria and yeast, the expression of many genes are triggered by changing environmental conditions. The genes devoted to responding to such environmental cues and achieving certain metabolic goals are often arranged in clusters known as *operons*.

Operons contain many genes. The transcription of these genes begins at a specific DNA sequence known as the ***promoter***, which is a binding site for the enzyme RNA polymerase. Binding of the RNA polymerase to the promoter initiates transcription. The transcription rate can be modulated by the action of suppressors and activators. Binding of suppressor molecules to the DNA sequences known as operators prevents RNA polymerase from carrying out transcription. An operator is strategically located between the promoter and the transcription initiation site of an operon. Activators often alter the binding properties of the suppressors to the operator.

Lactose operon in *E.coli* is one of the best studied examples of coordinate gene expression in prokaryotes. The operon consists of three genes, *lac Z*, *lac Y*, and *lac A*, encoding, respectively, β-galactosidase, lactose permease, and a transacetylase. The first enzyme hydrolyses lactose while the second protein helps lactose transport across the bacterial cell membrane. The function of the transacetylase enzyme is unknown. When glucose is available in the growth medium, bacteria do not produce any of these above enzymes. However, when only lactose is available as a carbon source these genes are expressed, allowing bacteria to metabolize lactose. This timely expression of genes is achieved by the presence of a promoter and an operator at the beginning of the *lac* operon. When glucose is plentiful, *lac* suppressor (the negative regulator of the *lac* operon), which is the product of the *lac I* gene, binds to the operator region and prevents transcription of the genes in the *lac* operon. Lactose can bind to the *lac* suppressor and inhibit its binding to the operator, relieving the inhibition of *lac* gene transcription.

The tryptophan operon consists of five genes which code for five enzymes in the biosynthetic pathway of tryptophan (Trp). The expression of these genes is downregulated when adequate amounts of Trp are available and upregulated when Trp is in short supply. When Trp is plentiful in the cytoplasm, it binds and activates suppressor molecules which can then bind to the operator and inhibit transcription. Here, Trp acts as a negative regulator. When Trp availability is low, the suppressor molecules are inactive and unable to bind to the operator, thus permitting the expression of enzymes needed for *de novo* synthesis of Trp.

There is an additional regulatory strategy employed in the tryptophan operon. A DNA sequence at the beginning of the operon is transcribed into a leader RNA molecule which can form a transcription inhibiting hairpin loop. This leader mRNA contains two consecutive Trp codons where ribosomes get temporarily stalled during translation when enough Trp is not available. This stalling prevents the formation of a transcription inhibitory hairpin loop.

In bacteria, genes are closely spaced in the chromosome. Selective expression of these genes requires that transcription is terminated at the end of the desired gene or operon. *E. coli* possess two mechanisms for ensuring transcription termination. The first requires the protein factor ρ which binds the nascent RNA and causes termination in an unknown manner. The second mechanism does not require ρ; hence, it is named ρ-independent transcription termination. This often occurs within a few hundred bases of the transcription initiation site at sites that carry characteristic sequences that form hairpin loops (**attenuators**), as seen in the Trp operon.

Prokaryotes have only one type of RNA polymerase. This enzyme consists of five subunits which interact with the promoter. Promoter sequences are not identical in all bacteria. The σ subunit of RNA polymerase that acts as an initiation factor may vary with the type of the promoter being used for transcription. The σ^{70} is the most common one.

Eukaryotic genes have more complex structures than prokaryotic ones. A eukaryotic gene consists of exons, introns, and transcription regulatory elements such as the TATA box, enhancers, and upstream promoter elements. Transcription factors bind to regulatory elements to aid transcription. These multiple regulatory elements and transcription factors allow for the complex and precise control of eukaryotic gene expression.

Cis-acting elements are DNA sequences located upstream of transcription start sites. They regulate the rate of transcription of downstream genes by binding various proteins. TATA box is such a *cis*-acting element, found upstream of rapidly transcribed genes that encode histones, globins, and other proteins. It helps to position RNA polymerase for transcription initiation. Other genes contain initiator elements instead of a TATA box. These initiator sequences are generally more degenerate than the TATA box.

Eukaryotic DNA is tightly folded with the help of histones and nonhistone proteins. Such tightly folded DNA is known as heterochromatin. Compared with naked DNA, heterochromatin is relatively resistant to **nucleases** like DNase I. Genes within heterochromatin are prevented from transcription due to their limited access to transcription factors.

Eukaryotes have three major RNA polymerases. They have complex subunit structures, with their larger subunits bearing homology to bacterial RNA polymerase. Each polymerase has enzyme-specific subunits whose functions are essential for viability but are poorly understood.

RNA polymerase II transcription initiation complex is assembled by adding many proteins in a specific order. These proteins are general transcription factors which activate transcription by helping the assembly of initiation complex. There are other proteins, however, that can inhibit this process.

Processing, transport, and translation of the RNA are also subject to specific regulation in eukaryotes. *Drosophila* sex determination illustrates how differential RNA processing can regulate gene expression. In *Drosophila* female embryos, sexlethal (Sxl) primary RNA transcript is spliced to exclude exon 3, which contains a stop codon. This leads to the expression of the functional Sxl protein that is needed for female differentiation of the embryo. In male embryos, exon 3 in the Sxl mRNA is never excised, resulting in premature termination of translation. In the absence of the Sxl protein, embryos develop as males.

Ferritin is an iron-binding protein. Its expression is regulated at the translation level in the following manner: Under conditions where the free iron level is low, iron response element binding protein (IRE-BP) binds to the 5' end of the ferritin mRNA and inhibits translation. When free iron is plentiful, it binds to IRE-BP and inactivates it, removing the inhibition to translation. The ferritin needed to store excess iron is then produced.

Self-Testing Questions

1. For your senior thesis project, you decide to isolate lac operon mutants. You grow wild type *E. coli* and mutagenize them with a potent chemical mutagen. Then you plate them in a medium containing x-gal and glucose.
   (a) What is the purpose of adding x-gal into the medium?
   (b) What would be the color of the wild type *E. coli* colonies be if you were to add IPTG, a *lac* operon **inducer**, into the medium? Why?
   (c) The majority of the colonies grown on the plates without IPTG were white, but a few were blue in color. List the types of mutations that can give rise to constitutive expression of β-galactosidase.
   (d) Classify the mutant genetic elements you proposed in part (c) as *cis*-acting or trans-acting.
   (e) Mutants of the *lac* operator region, **constitutively expressing** β-galactosidase, were further mutagenized and screened for revertants of the blue phenotype to white phenotype. There were two distinct classes of revertants. Class A mutants produced blue colonies in the presence of IPTG or galactose while the Class B mutants did not. Explain this observation by making suggestions for the additional elements that may have been mutagenized in the second mutagenesis procedure.
   (f) A plasmid was constructed that contains all the elements of the *lac* operon except for the *lac I*. This construct was transfected into the class A and class B mutants in part (e). Predict the phenotype of these mutants after transfection in the presence and absence of IPTG.

2. You have cloned a novel gene X that is believed to play a central role in the pathogenesis of a rare genetic disease. You decide to study how the expression of this gene is regulated in order to understand the physiological role of the protein encoded by X. You have the DNA sequence for the 300 bases upstream of X starting from the +1 transcription start site.

(a) If X were to be expressed at high levels, what type of genetic elements would you expect to see in the upstream 300 bases?

(b) It turns out that the region upstream of X does not contain any of the classic elements that are known to elevate levels of expression. What would be the characteristic of the composition of bases that would indicate the presence of a promoter?

(c) In order to identify the exact upstream promoter region, you prepare DNA segments of various length from the upstream sequence by using PCR. Then you link them to a downstream β-galactosidase gene that acts as a reporter gene. These constructs are then integrated into a DNA vector and transfected into cultured mammalian cells, and the expression of β-galactosidase is then measured. Draw a cartoon to illustrate the arrangement of the different DNA elements (i.e., upstream sequence segment of the gene X, β-galactosidase gene, vector backbone) in the vector that you transfected into mammalian cells.

(d) How does this transfection experiment help you identify the hidden promoter element located upstream of gene X?

(e) How would you measure the level of expression of β-galactosidase?

(f) Why are cultured mammalian cells used for these transfections instead of bacterial cells?

Indicate whether the following statements are True or False.

3. A mutant bacterium has a mutation in the *lac I* gene. A mutant protein that is unable to bind lactose, but otherwise normal, is being produced. This mutant will have constitutive expression of the lac operon genes.

4. The catabolite activator protein (CAP) bound to cAMP is essential for high expression of lac operon genes.

5. Lac operator binds the CAP-cAMP complex.

6. Deletion of the sequence 1 of the leader sequence of the tryptophane operon will lead to hairpin loop formation between the sequences 2 and 3 of the leader mRNA. This mutant bacteria will constitutively express Trp operon genes because the attenuator hairpin loop is never formed.

7. Extremely rapid expression of heat shock proteins is triggered by prior assembly of the transcription initiation apparatus, which is the result of the heat shock transcription factor being activated by stress.

8. Sigma factor is a protein that helps initiation of transcription by RNA polymerase in bacteria. There are multiple sigma factors that help RNA polymerase recognize different promoters.

9. HIV Tat protein interacts with cellular proteins and RNA strands that are being synthesized to form a loop of RNA. This RNA loop interacts with RNA polymerase to prevent premature termination of the transcription.

10. Transposition of a eukaryotic gene by the activity of a mobile genetic element (also known as a transposon) to a region of heterochromatin may lead to the repression of the gene.

11. The inactive X chromosome in mammalian females mostly contains genes that are repressed. However, there are some genes on the inactive X chromosome that escape this "X inactivation."

12. Most of the mRNA synthesis occurs in the nucleolus of eukaryotic nuclei.

Fill in the blanks of the following sentences to make an accurate statement.

13. Prokaryotic RNA molecules that carry the coding information for multiple proteins are called _____ mRNA.

14. Rapidly transcribed eukaryotic genes contain the cis-acting regulatory elements called the _____ at the 26-34 bases upstream of the transcription start site.

15. Eukaryotic DNA is tightly folded and packed inside the nucleus with the help of basic proteins known as _____. These proteins are particularly rich in amino acids _____ and _____.

16. The nuclear DNA that is transcriptionally inactive and resistant to digestion by DNase I is known as _____. The transcriptionally active DNA that looks pale in electron micrographs is known as _____.

17. In prokaryotes, one type of transcription termination is dependent on a protein factor called _____.

18. Processing of tyrosine tRNA involves four types of changes, which are _____, cleaving off of a 5'-end sequence, _____, and adding an ACC sequence to the 3'-end.

19. The expression of some eukaryotic genes are regulated by controlling the translation rate. _____ and _____ are some examples of proteins whose translation is regulated by the protein IRE-BP.

20. In _____ Drosophila embryos, the primary RNA transcript of the sex lethal gene (Sxl) is spliced to exclude the exon 3, which contains a stop codon. This is a mechanism by which the sex of a developing embryo is determined.

21. The λ bacteriophage exhibits two different patterns of life cycles. They are known as _____ and _____ life cycles.

22. A new bacteriophage λ infecting a bacterium containing a dormant λ prophage will not undergo multiplication. This is because of the cooperative binding of the _____ protein to the $O_R 1$ and $O_R 2$ operators.

Answers to Self-Testing Questions

1. Cellular Chemistry

1. (a) (i), (ii), (iii), (iv), (ix), (x)
   (b) (v), (vi), (vii), (viii), (xi), (xii)
   (c) (iv)
   (d) (xi), (xii)
   (e) (i), (ii), (ix). Part (a) also contains molecules that bear electrical charges.
   (f) (vii) and (viii) as well as (xi) and (xii) are structural isomers.

2. Melting points in descending order.
   $$(iii) > (iv) > (ii) > (i)$$
   The larger the molecule, the stronger the van der Waal's interaction. However, the trans isomer hinders tight interactions between molecules and the melting point is therefore lower.

3. (a) Point mutations may affect all levels of the structural organization of a protein. This depends; however, on the location of the point mutation in the primary structure and the nature of the substitution. If the substituted amino acid is very similar, nothing more than the primary structure may be affected.
   (b) The mutation in A must have changed the tertiary structure of A so that its interaction with wild type B is disrupted. The revertant has a mutation in protein B that changes its tertiary structure so that it can once more interact with A.

4. True

5. False. Triglycerides made of saturated fatty acids solidify more easily than triglycerides containing unsaturated fatty acids because the rigid trans configuration around the double bonds favors interaction between molecules.

6. True

7. True

228   Answers to Self-Testing Questions

8. True
9. False. Fructose is a monosaccharide found in fruit.
10. False. Lecithin is one of the major phospholipids found in cell membranes.
11. True
12. False. Steroids contain four interlocking carbon rings. A phospholipid molecule consists of two long chain fatty acids esterified to glycerol which in turn is attached to a phosphate group which carries an amino alcohol like choline.
13. True
14. glucose; fructose
15. glycogen
16. glycerol; phosphate; amino alcohol (like choline)
17. amino acid sequence
18. alpha helix; beta pleated sheet
19. B
20. E
21. F
22. H
23. I
24. G
25. M
26. C
27. A
28. O

2. Protein Structure and Function

1. Amino acids with hydrophobic R groups make up most of the membrane spanning helices. Check out the following example, glycophorin, which is a transmembrane protein found in the membranes of red blood cells. The majority of the amino acids embedded in the membrane consist of Leu, Ile, Val, Met, Phe and Ala, all of which contain hydrophobic R groups. It is

believed that the charged amino acids found in the membrane interact with each other and neutralize their charges, enabling them to interact with hydrophobic membrane lipids.

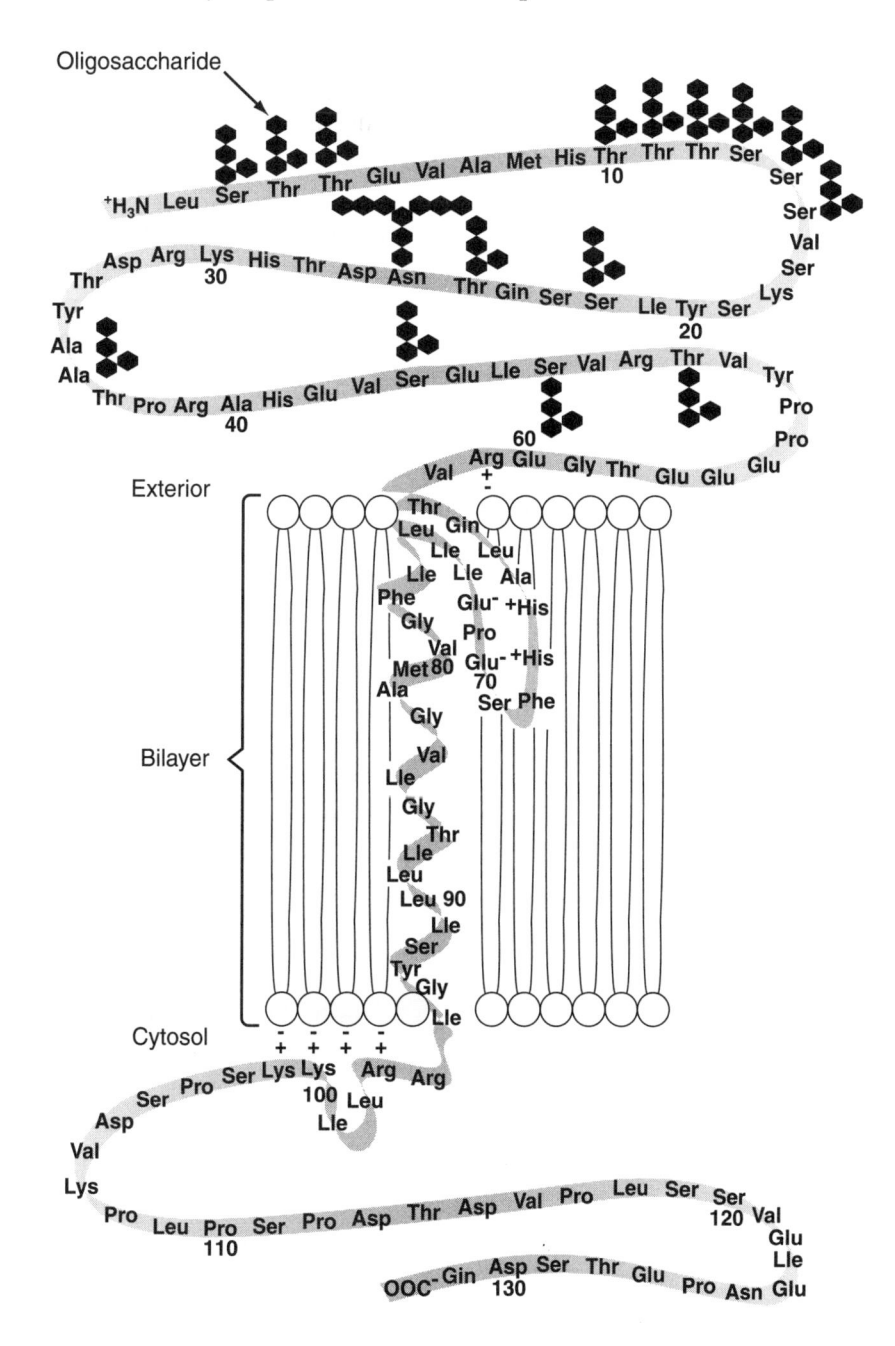

2. (a) Depending on the concentration, salt destabilizes non-covalent interactions such as ionic interactions and hydrogen bonds. At one concentration, salt may destabilize the interactions between the subunits of the protein and may affect the quaternary structure. At another concentration, the salt may disturb the interaction of the protein with water and make proteins aggregate and precipitate.

   (b) Extremes of pH affect the ionic charges borne by the R groups of amino acids such as arginine, lysine, glutamic acid and aspartic acid. These changes affect the interactions that may determine the secondary, tertiary and quaternary structures of the protein, resulting in a denatured protein.

   (c) Reducing agents disrupt the disulfide bonds between cysteins, thereby disrupting the tertiary structure of the protein. Proteases are able to break peptide bonds at specific sites of the protein's primary structure. These protease cleavage sites may be embedded in the tertiary structure of the folded proteins. Prior denaturation of hemoglobin in this experiment exposes these sites to protease cleavage.

   (d) Sickle cell anemia is a result of a single base substitution in the DNA encoding the β chain of the hemoglobin. This base substitution replaces glutamic acid (which is a charged amino acid) at the sixth position with valine (a neutral amino acid). The abnormal polymerization of sickle hemoglobin at low oxygen tensions occurs as a result of this mutation. When the mutant protein and the normal wild type protein were digested with a mixture of proteases, both give rise to the same combination of peptides with the one exception of the fragment containing the mutated amino acid. Since valine does not bear a charge at physiological pH and glutamic acid does, the two peptides exhibit different electrophoretic mobility.

3. (a) Following are some of the possible mutations:
   - Mutant lysosomal M6P receptors are unable to bind lysosomal proteins with phosphorylated

mannose residues attached to them. The interaction between M6P and its receptor is a fundamental mechanism that sorts lysosomal enzymes to the right compartment. Mutant M6P can explain the phenotype of multiple lysosomal proteins being aberrantly sorted.

- There may be a deficient enzyme that affects adding phosphates to the right mannose residues.
- There may be a mutation in the enzymes involved in adding sugar groups to lysosomal proteins. This may affect processing of the newly synthesized proteins through glycosylations and subsequent modifications.

(b) The patients with I-cell disease appear to have functional M6P receptors, as they are able to take up wild type lysosomal proteins from the culture medium. They are also able to take up their own lysosomal enzymes once they are phosphorylated. This indicates that the defect is in the phosphorylation machinery.

(c) As in part (b), the M6P receptors on the cell surface are able to take up processed enzyme from the medium. If you infuse the missing enzyme into patients with Gaucher's disease, the macrophages in their reticuloendothelial system are able to take up the enzyme, and the enzyme becomes functional in the lysosomes. This strategy, in fact, is used in treating patients with Gaucher's disease. Wild type enzyme is isolated from the placentae of normal newborns and processed to expose M6P residues, then injected into Gaucher's patients periodically.

4. (a) Eukaryotic proteins recombinantly expressed in bacteria may not fold properly for many reasons. Some of them are listed below.

- Bacterial cytoplasm is too reducing an environment for the formation of disulfide bonds; any eukaryotic protein expressed there may not form the appropriate disulfide bonds.

- Bacteria do not contain any membrane-bound organelles like the endoplasmic reticulum (ER) and Golgi apparatus. For some eukaryotic proteins, processing in these compartments is essential in order to attain their final proper conformation.
- Same chaperone proteins present in eukaryotes may not be present in the bacterial cytoplasm.
- Folding signals contained in the protein may not be recognized by the bacterial protein folding machinery.
- Bacterial proteases present in the cytoplasm may degrade the foreign, misfolded eukaryotic proteins.

(b) This protein appears to belong to the class of chaperone proteins. It may bind to unfolded protein in the cytoplasm and remain tightly associated with presenillin-1. When ATP and Mg^{++} are added, hydrolysis of ATP releases the chaperone from presenillin-1. (Do you remember how Bip worked?)

(c) Bacterial cytoplasm is too reducing for disulfide bond formation to occur. Most of the cysteins stay as -SH rather than forming disulfide bonds (-S-S-).

5. False. Peptide bonds join two adjacent amino acids of a polypeptide.
6. True
7. False. There are some RNA molecules that exhibit enzymatic activity.
8. False. Glycoproteins acquire the carbohydrate side chains after the protein is synthesized, during the processing in the endoplasmic reticulum and the Golgi apparatus.
9. True. Proinsulin and trypsinogen are two examples.
10. False. There are many small peptides such as enkephalins, endothelins, and angiotensin, that perform important biological functions.
11. True
12. False. GPI anchors help proteins localized to the plasma membrane.

13. True

14. True

15. quartenary

16. chaperons

17. collagen

18. allosteric

19. structural; cytoskeleton

20. hydrophobic; hydrophilic

21. hydrophobic

22. peptide; amino acids; primary

23. R

24. domains

3. Nucleic Acids and Protein Synthesis

1. (a) Primer 1: 5'-CCACCATGGGGAGTCTCAGCC-3'

    Primer 2: 5'-CCTCACACCCAGGAGTCGGGGG-3'

    Notice that both primer sequences are written from the 5' to 3' direction, the direction of DNA growth. Effective primers are roughly 20 bases in length, have more than 50% G/C content, and have high levels of G/C at the ends of the primer. All these factors help in annealing the primer to the template strand.

    (b) Four nucleoside triphosphates (dATP, dGTP, dCTP, dTCP), DNA polymerase, and a buffer to maintain proper pH.

    (c) The α phosphate of the NTP needs to be labeled with radioactive phosphate. The β and γ phosphates are removed during the incorporation of the nucleotide.

2. (a) The strand complimentary to the given DNA strand will be used as the template for transcription.

    (b) An open reading frame is the base sequence that gives a coding region uninterrupted by stop codons. In the given sequence an open reading frame is found from the first ATG. The amino acid sequence is

5'Met-Gly-Ser-Leu-Ser-Gln-Ser ... Ser-Pro-Asp-Ser-Trp-Val-STOP.

(c) This will introduce a stop codon (UAG) at the site and interrupt the open reading frame.

(d) A Kozak sequence is found at the translation start site of mRNA. This acts as a signal for the ribosomes to start translation. A Kozak sequence at the 5' end of the given sequence can be identified (5'-ACCAUGG-).

(e) Shine-Dalgarno sequences help recruit small ribosomal subunits to mRNA in *E. coli*. This sequence is complementary to a sequence at the 3' end of the 16S small ribosomal RNA. The consensus sequence of Shine-Dalgarno sequence is as follows.

mRNA 5'-UAAGGAGG—(5–10 nucleotides)-AUG(translation start site).

(f) In the given sequence, a Shine-Dalgarno sequence cannot be found, because not enough sequence upstream of the translation start site is given.

3. True

4. False. Glycerol is not a component found in nucleotides. A sugar (ribose or deoxyribose), a phosphate group and a nitrogenous base (purines or pyrimidines) make up a nucleotide. Glycerol is a component of phospholipids.

5. False. Bacterial genes do not contain intervening sequences. Eukaryotic genes do.

6. True

7. True. This is done in a PCR, 20–40 times.

8. True

9. True

10. True

11. False. There are only minor differences in a few codons between bacteria and eukaryotes. For this reason, the genetic code is said to be universal. Many antibiotics inhibit bacterial protein synthesis by specifically binding to bacterial ribosomes, which are different from eukaryotic ribosomes.

Answers to Self-Testing Questions   235

12. True. This is found in eukaryotic genes.
13. nucleotides
14. phosphodiester
15. gene
16. operons
17. endonuclease
18. introns, exons
19. semiconservative
20. 5' - 3', Okazaki fragments
21. exons, methyl-guanosine cap, poly-A tail
22. Shine-Dalgarno

4. Techniques

1. Abbe's
2. phase, confocal, dark-field, and Nomarski (differential interference contrast)
3. fluorescence
4. confocal
5. freeze-fracture or freeze-etch
6. nucleic acids
7. RNA
8. polymerase chain reaction
9. cDNA
10. size
11. isoelectric focusing; SDS gel electrophoresis
12. antibodies
13. autoradiography
14. (b)
15. (c)
16. (b)
17. (d)
18. (e)

19. (b)
20. (b)
21. (e)
22. (b)
23. (c)
24. (a)
25. (d)
26. Paraffin embedding provides a hard matrix so that tissue can be sectioned and stained. Alcohol replaces the water, xylene replaces the alcohol, and liquid paraffin replaces the xylene. This series is necessary to replace the water with the solid matrix paraffin.
27. The dark-field microscope causes the bacteria to be bright against a dark background, thus increasing their contrast. The phase microscope, however, is simply not as good at increasing the contrast of bacteria.
28. It shows the approximate surface of the cell, which, in turn, facilitates insertion of the micropipet.
29. **Bright-field and phase**   **Confocal**
    Light   Laser
    No computer   Computer necessary
    Entire image in field   Images are points
    Relatively cheap ($10,000)   Expensive ($200,000)

    The confocal microscope has allowed us to increase the practical limit of resolution of microscope systems and to optically section specimens. We can also construct 3D images not obtainable before the 1980s.
30. Fluorescent probes have a rich diversity in potential. For instance, they can be linked to antibodies for antigen localization, measure membrane fluidity, monitor changes in intracellular function, and directly stain organelles and structures like DNA. Also, many fluorescent probes have been developed to replace radioisotopes, which pose more of a threat to the environment.
31. In photobleaching, a strong light source, usually a laser, causes fluorescent molecules to fluoresce so that they

Answers to Self-Testing Questions 237

become "bleached." This technique can be used to monitor plasma membrane fluidity through a process called FRAP (fluorescence recovery after photobleaching).

32. (a) antigen: any molecule that elicits an immune response; antibody: bivalent or multivalent molecule that has two binding sites for the same epitope.

    (b) rhodamine and fluorescein: two fluorescent molecules that fluoresce red and green, respectively.

    (c) direct, indirect: two different techniques of accomplishing immunocytochemistry or ELISA assays. The former employs one antibody, while the latter uses two antibodies.

    (d) monoclonal: one antibody from one B cell; polyclonal: many antibodies from many B cells.

33. (a) Cells are labeled with a "pulse" of radioactivity followed by a "chase" of nonradioactive isotope.

    (b) Two different labels to show the presence and association of two types of structures or molecules.

    (c) Antigens are precipitated out of solution using antibodies attached to protein A, which is attached to a biologically inert bead.

    (d) Antibodies localize a protein on a gel following transfer to nitrocellulose paper.

    (e) This technique separates proteins using a ligand-receptor interaction.

34. (a) The polyclonal approach is easier but has a higher probability of cross-reacting with similar proteins due to the diversity in antibodies produced by many B cells. Also, the production of the antibody is limited by the life of the donor. A monoclonal antibody, on the other hand, is immortal because it is produced by cultured cells. It is more difficult to make and has less chance of cross-reacting, however.

    (b) The direct approach is difficult because it is necessary to covalently link fluorochrome with antibody. With the indirect technique, the secondary antibody can be bought directly from a commercial supply house.

238 Answers to Self-Testing Questions

(c) With transmission electron microscopy the internal structure of cells is visible, whereas with scanning electron microscopy only the surface structure is seen.

35. (a) gel electrophoresis, preferably with ^{35}S-methionine labeling and autoradiography

(b) protein A immunoprecipitation, Western blot, ELISA

(c) affinity chromatography

(d) protein A immunoprecipitation or affinity chromatography

(e) *in situ* hybridization

(f) sequencing

(g) ultrastructural immunocytochemistry

(h) SDS or native gel electrophoresis

(i) fluorescence microscopy

(j) light microscopy autoradiography using ^{3}H-uridine

(k) microinjection of lucifer yellow or similar dye using intracellular pipet

(l) fluorescence microscopy and suitable fluorescent calcium indicator dye

(m) intracellular injection (iontophoresis only) with lucifer yellow or some similar dye

(n) electroporation or liposomes

36. (a) SDS gel electrophoresis can separate multimeric proteins into their constituent parts. The 200-kD protein appears to be a dimer consisting of two units of 100 kD, both of which co-migrate in the gel and appear as one band. Thus, on the SDS gel the protein appears as 100 kD, whereas on the native gel it appears as 200 kD.

(b) The glutaraldehyde cross-linked the protein so that the beta-mercaptoethanol and urea used in SDS gel electrophoresis can no longer separate the dimer into two monomeric units.

(c) One possibility might be that the aldehyde has changed the shape of the protein, thus changing the epitope. Therefore, the antibody doesn't bind.

(d) One possibility is that you forgot to permeabilize the cells so the antibody could get inside the cell to seek out its epitope. Other answers are possible.

(e) Do differential centrifugation (velocity sedimentation) to separate nuclei from cytoplasm. Take the pellet containing the nuclei, homogenize it, and look for the 200-kD element, using an ELISA assay. Do the same for the supernatant (cytoplasm).

(f) Pulse-chase-label cells with ^{35}S-methionine. Immediately immunoprecipitate with protein A and run on a gel followed by SDS gel electrophoresis. The result should be one band with a density that can be quantified on a densitometer. Do the same experiment again, but wait for a certain time (hours to days). The time it takes the radioactive band to reach half the intensity it was originally is the half-life of the protein.

37. Pre-absorb the antibody with antigen. This should prevent all binding. If binding still occurs, then nonspecific binding is probably the problem. Another idea would be to decrease the concentration of antibody being used. At high concentrations (1:10) antibodies tend to stick nonspecifically, whereas at low concentrations (1:100 to 1:500) they have less of a tendency to do this.

38. A cell strain is mortal; a cell line is immortal.

39. There are many possibilities including Southerns, Northerns, subtractive hybridization, *in situ* hybridization, stringency tests for comparing evolution of species, and genetic testing.

40. DNA fragments tend to align themselves with the electric field during separation—an attribute not shared by proteins. The pulsed field allows DNA to realign perpendicular to the electric field so that better separation is achieved.

41. Westerns identify proteins, Southerns identify DNA sequences, and Northerns identify RNA. All use gel electrophoresis and transfer paper.

42. The basis of subtractive hybridization is to use two similar cells that contain similar genes, except for the one of interest. By combining the DNA from both cells subsequent to denaturation, one can separate the unpaired DNA, which codes for unique genes.

43. One can combine single-stranded DNA of the same gene from two different species. The greater the temperature needed to "melt" (denature) the DNA, the fewer the mismatches that exist and the closer the genes are in evolution.

44. Most enzymes denature at 100°C, but the DNA polymerase from *T. aquaticus* is still functional at this temperature. This high temperature is necessary for the proper dissociation of the two complementary strands.

5. Cell Membranes

1. E
2. G
3. B
4. D
5. C
6. A
7. F
8. amphipathic
9. negative
10. detergents; salt
11. *cis*
12. Cholesterol
13. phase
14. hydrophilic
15. light; protons
16. purple; *Halobacterium*
17. Gangliosides
18. fluid-mosaic
19. spectrin; band 3

Answers to Self-Testing Questions 241

20. apical; basal
21. Lanthanum hydroxide
22. Nernst
23. D-; L-; D-
24. Ouabain
25. antiport
26. Freeze-fracture or freeze-etch
27. higher
28. (c)
29. (a)
30. (e)
31. (e)
32. (c)
33. (b)
34. (d)
35. (a)
36. (c)
37. (b)
38. (b)
39. (d)
40. (e)
41. There are many possible answers to this question. One protocol is the following: Erythrocytes could be harvested, and inside-out and right side-out ghosts could be developed. Next, the vectorial label could be added to both these preparations. Once the radioactively tagged label has covalently linked itself to the exposed portion of the integral membrane protein, the ghosts could be solubilized in a biological detergent such as sodium dodecylsulfate, and the proteins separated on SDS gels. Next, an autoradiogram of the proteins on the gel could be generated so that the proteins that are radioactively labeled can be identified. By comparing the gel of the inside-out ghosts with the gel of the right

side-out ghosts, the topography of the proteins in question should be clear.

42. Lectins can bind to the carbohydrate groups on glycoproteins and glycolipids. These carbohydrate groups are predominantly found on the external surface of cells. Thus, it would be possible to link lectins to a bead in a column to use as an affinity sorting tool. Presumed mixed populations of ghosts could then be equilibrated on the column. The right side-out ghosts would be retained by the column owing to the presence of the carbohydrate moieties and their association with this lectin, while the inside-out ghosts would be easily purified because they would not be retained by the affinity matrix.

43. Spectrin probably associates with ankyrin and possibly other spectrin molecules, forming an aggregate with an apparent, combined molecular weight in excess of that of the spectrin tetramer.

44. Phenol red is a pH indicator dye. It turns yellow in the presence of excess protons (acidic conditions) and turns purple in the presence of hydroxyl ions (basic conditions). Bacteriorhodopsin could be placed in a bathing medium that has phenol red, and then the bacteria could be subjected to a condition that either increased or decreased proton pumping. The media from the control and the experimentals could then be analyzed using a spectrophotometer. By using an appropriate standard curve using phenol red and known acids or bases, a direct estimate of the proton-pumping activity of the bacteriorhodopsin molecule could be estimated.

45. There are a number of possible approaches that could be used to answer this question. You could prepare liposomes, artifical phospholipid bilayers with no proteins, and accomplish freeze-fracture using the suspect protocol. If the large particles appear, then they clearly are not integral membrane proteins and are merely artifacts of the experimental protocol. If they don't appear in the liposome replicas, they still might be an artifact in your specimen membranes, but at least some possibilities have been ruled out.

Answers to Self-Testing Questions

46. (a) One possibility is that the antibody is bound to proteins that are connected to the inner cytoskeleton via actin filaments. If this is the case, these proteins will have a much reduced mobility. Indeed, nearly half of all transmembrane proteins in cells are linked in this manner. To test this hypothesis, cells could be pretreated with cytochalasin, a drug that depolymerizes actin filaments. If the proteins now cap in the presence of this drug, you can assume that capping did not occur in the untreated cells because actin restricted the ability of the protein to move laterally within the plane of the membrane.

    (b) Native antibodies are bivalent and, as such, can bind to two epitopes (two membrane proteins) at the same time. This allows the antibodies to precipitate the proteins in a patch. A monovalent antibody can bind to only one cell membrane protein and thus cannot form the connections between proteins required to create the clustered mass of proteins that makes up the patch.

47. First, radioactively tag the molecule so that its transport kinetics can be measured. Second, measure its transport rate in the presence and then the absence of an ATP inhibitor such as oligomycin. If the rate is reduced when ATP is reduced in the cell, then active transport should be suspected. If, on the other hand, the rate is not significantly altered, then facilitated transport should be considered.

48. The ability of ion carriers to transport their respective molecules is very sensitive to cell membrane temperature, whereas this is not the case with channel formers. When the cell membrane cools, the plasma membrane undergoes a phase transition during which the diffusion characteristics change substantially. Under normal physiological conditions, such as those that might occur in a cold-blooded animal subjected to cold, different phospholipids and cholesterol concentrations are changed to maintain fluidity. But under the *in vitro* conditions described in the question, no such changes in membrane constituency occur. Thus an ion carrier's behavior and a channel former's transport characteristics are different as exhibited by the following figure.

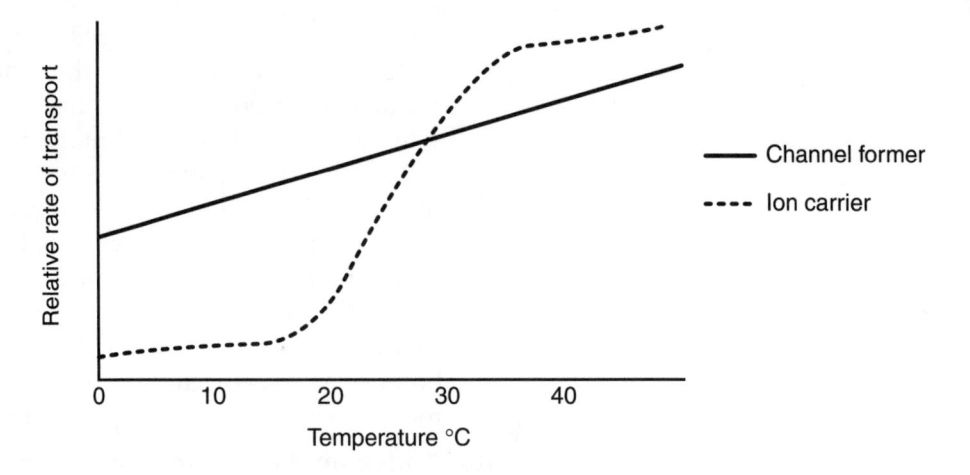

By comparing the transport characteristics of this new ionophore, you should be able to determine if it is an ion carrier or a channel former.

49. There are more hydrogen bonds between the phospholipid molecules and the transmembrane protein when the protein is in the α helix form rather than the β pleated sheet. This increases the stability of the protein in the membrane.

50. One solution is to grow a monolayer of the kidney cells on a filter. Next, add sodium fluorescein to the apical (top) portion of the cells and then at intervals thereafter collect samples from the basal (bottom) compartment. The amount of sodium fluorescein that leaks from the apical compartment to the basal compartment is a direct measure of the tight junction's ability to maintain separation of the apical and basal media.

6. Protein and Vesicular Traffic

1. Nonvesicular
2. Vesicular
3. Nonvesicular
4. Nonvesicular
5. Nonvesicular

Answers to Self-Testing Questions 245

6. Vesicular
7. Nonvesicular
8. Nonvesicular
9. Vesicular
10. Vesicular
11. resident
12. Flippase
13. Dolichol phosphate
14. *cis*
15. acidic
16. regulated
17. chase
18. pores
19. signal peptidase
20. glycosylphosphatidylinositol
21. (b)
22. (d)
23. (a)
24. (d)
25. (b)
26. (a)
27. (c)
28. (e)
29. (e)
30. (e)
31. (a) Brefeldin A blocks the ability of transport elements (vesicles) to be transported from the rough endoplasmic reticulum (RER) to the Golgi. So, secretory proteins will not move forward in their pathway after synthesis in the RER.

    (b) Nocodazole is a microtubule inhibitor. Returning vesicles from the Golgi to the RER containing resident RER enzymes and their receptors depend on microtubules. When the microtubules are disrupted, these enzymes and their associated receptors are not returned to the RER.

(c) EGTA decreases the free extracellular calcium. Release of secreted protein through the regulated pathway can be governed by an influx of calcium subsequent to, for instance, a ligand-induced stimulus. EGTA-AM, on the other hand, can cross the membrane and decrease intracellular calcium. So, the addition of EGTA and EGTA-AM to cells undergoing this process could cause proteins to be retained in storage vesicles.

(d) Protein synthesis inhibitors, such as puromycin, have an immediate effect on proteins secreted through the constitutive pathway whereas there is no immediate impact on proteins being released through the regulatory pathway. This is because proteins being secreted through the regulatory pathway are stored, and the influence of protein synthesis inhibitors is not felt until later. Proteins secreted through the constitutive pathway are not stored, so their secretion is directly tied to protein synthesis.

(e) The release of lysosomal enzymes from their mannose 6-phosphate (M6P) receptors in the lysosomes is contingent on the acidity of the endosome. Ammonium chloride and chloroquine can increase the pH to near neutrality and prevent this dissociation. As a result, fewer lysosomal enzymes are deposited in the lysosome because many return from the lysosome to the trans Golgi network.

32. Collagen is a secretory protein synthesized in the RER. It contains a signal sequence. The rough microsomes are the centrifuge-equivalent of RER and contain a signal peptidase that clips off the signal peptide from the collagen. Collagen, however, can also be synthesized *in vitro* on free ribosomes. When this occurs, no signal peptidase is present and the signal peptide isn't clipped off. The signal is about 24 amino acids long. Multiplying 24 times the average molecular weight of an amino acid (110) yields an approximate 2600 molecular weight.

Answers to Self-Testing Questions 247

33. Protease treatment of the microsomes cleaves the docking protein (SRP receptor) necessary for the proper binding of the SRP and its associated complex. Thus while collagen synthesis is initiated on free ribosomes, it is unable to continue in system A because the ribosome-mRNA SRP complex is unable to dock to the microsome.

34. If your protein is shipped via vesicular transport, much of the protein would be encapsulated in secretory vesicles as well as vesicles that resealed from the fractionation of the rough endoplasmic reticulum, Golgi, and trans Golgi network. Pretreat the supernatant with protease and then look for the protein on a Western blot. If the protein is in a vesicle, it will be protected from the protease and appear intact. But if protease treatment of the supernatant yields a smear or no protein on the Western blot, you can conclude that this protein was not encapsulated in a vesicle; rather it was synthesized on free ribosomes like proteins destined to go to the nucleus, chloroplasts, mitochondria, and peroxisomes. Consequently it was vulnerable to protease attack.

35. (a) The protease probably clipped a receptor on the outer membrane that is necessary for the proper translocation of this protein.

    (b) The translocation of this nucleus-encoded mitochondrial protein does not depend on ATP synthesis. This can be concluded, since oligomycin inhibits ATP synthesis at the level of the F_0F_1 particle (see Topic 8 on Energy, Mitochondria, and Chloroplasts for details). DNP, however, decreases the proton gradient across the mitochondrial membranes. This proton gradient is known to be important to the translocation of mitochdondrial proteins targeting the matrix.

36. One approach is to use the radioactive isotope ^{3}H-fucose. Add the isotope to living cells, wait an hour or so, and then separate the RER from the Golgi using centrifugation techniques. Next, immunoprecipitate the protein in question and determine if it is radioactive by using scintillation counting. Radioactivity would sug-

gest that the ³H-fucose has been incorporated at that particular site.
37. Resident RER proteins all contain a KDEL sequence that is recognized by the KDEL receptor.
38. Add labeled lysosomal enzyme to the extracellular medium surrounding the cells in culture. M6P receptors on the outside of cells can endocytose lysosomal enzymes that are mistakenly secreted by the cell through the default pathway. If the receptor is defective, no internalization of the isotope-labeled enzyme will be detected. This can be assessed using a scintillation counter.
39. The probable reason underlying this phenomenon is that the cytosolic fraction included chaperonins, the ATP-dependent chaperone proteins (such as hsp70) that are necessary for translocation because they maintain translocating mitochondrial proteins in an unfolded array.

7. Receptors and Second Messengers

1. E
2. B
3. A
4. D
5. C
6. cAMP-dependent phosphodiesterase
7. GDP; GTP
8. K_D
9. Cholera toxin
10. regulatory; catalytic
11. **SH2**
12. calcium
13. glucose
14. breakdown (phosphorolysis)
15. catalytic

Answers to Self-Testing Questions   249

16. (b)
17. (c)
18. (e)
19. (a)
20. (c)
21. (e)
22. (b)
23. (b)
24. (a)
25. (d)
26. (a)
27. (c)
28. (b)
29. (e)
30. (a) A has the lower K_D. The actual values can be calculated by the following equation: $K_D = [R][L]/[RL]$, where [R] and [L] are concentrations of free receptor and hormone, respectively, and [RL] is the concentration of the receptor-hormone complex. The K_D value is equivalent to the concentration of hormone at which half the receptors are occupied with hormone (ligand).

    (b) The most probable answer is that A and B are binding to two different sites on the same receptor. Alternatively, one or both are binding nonspecifically to nonreceptors.

31. The cell line with the dopamine receptor might have a dopamine receptor linked to a G_i (inhibitory G) protein. The fused heterokaryon now has an epinephrine receptor that links to both a G_i and its original G_s (stimulatory G) protein. Thus, stimulation of the epinephrine receptor stimulates both G proteins, resulting in only a modest increase in cAMP.

32. Add 8-azido-cAMP^{32} to the cell preparation, irradiate and separate on an SDS gel. Determine molecular weights by doing autoradiography using X-ray film. The

film will enable you to localize the radioactive label and, hence, the regulatory unit or units to which it is bound.

33. GMP-PNP acts like GTP by stimulating the G protein. GTP, however, is normally hydrolyzed to GDP, thus inhibiting the ability of adenylate cyclase to synthesize cAMP from ATP. GMP-PNP can't be hydrolyzed, and thus the system is irrevocably turned on. Cholera toxin inhibits the ability of GTP to be hydrolyzed to GDP, causing a similar effect of increasing the cAMP accumulation up to 100 times more than normal. Since both agents work on the same G protein through the same GTP mechanism, the combined effect of both is usually not larger than either alone (assuming that each is used at saturating concentrations).

34. (a) The easiest protocol would be to construct a regime by which insulin-dependent glucose transport can be measured. In this case, you could add ^{3}H-glucose to the outside of liver cells with and without insulin. If the cell transduction process is operating correctly, there should be increased internalization of the radioactive glucose in the presence of insulin. This could be assessed by homogenizing the cell and quantifying the radioactivity using a scintillation counter.

(b) Stimulate mutant cells with insulin in the presence of $^{32}PO_4$. Immunoprecipitate the insulin receptor from detergent-solublized cell homogenates. Separate on SDS gels and detect possible phosphorylation using autoradiography. Further analysis could determine which tyrosines are properly phosphorylated.

35. One mutant might be deficient in GAP (GTPase activating proteins), while the other might be defective in GEF (guanine nucleotide–exchange factor). [Note: GEF is also referred to as GNRP (guanine nucleotide–releasing protein)]. When the two mutants are combined into a heterokaryon, each mutant supplies the wild-type factor that is missing from the other mutant. Each mutant site is thus completed by a wild-type site.

Answers to Self-Testing Questions 251

36. A23187 causes an increase in intracellular calcium that mimics the release of calcium from the smooth endoplasmic reticulum—a necessary prerequisite for protein kinase C activity. The tumor promoter, on the other hand, binds to and activates protein kinase C because it is a chemical mimic of 1,2-diacylglycerol.

37. One way that you could approach the problem is to add dbcAMP to the cells possessing the epinephrine receptor. dbcAMP is a cAMP analog that, unlike native cAMP, is membrane-soluble. It can therefore activate cAMP-dependent protein kinase without having to have epinephrine bound to its receptor. If $^{32}PO_4$ is added, then these radioactively tagged cAMP-dependent phosphorylated proteins can be distinguished using SDS gel electrophoresis and autoradiography. Addition of the dbcAMP mimics epinephrine-stimulated cAMP stimulation but should not cause the phosphorylation of any proteins phosphorylated by non-cAMP pathways. By comparing an autoradiogram prepared from cytosolic extracts from dbcAMP-stimulated cells an autoradiogram prepared from cytosolic extracts from epinephrine-stimulated cells, it might be possible to determine if some of the phosphorylated proteins are being phosphorylated through mechanisms other than through cAMP.

38. You could use a calcium indicator dye such as Fluo3 or aequorin. Both of these dyes increase their intensity in response to a rise in intracellular calcium from the smooth endoplasmic reticulum. If the endoplasmic reticulum is releasing calcium, you should see transient "calcium spikes" that respond to the addition of the appropriate hormone.

39. One approach would be to add a protein synthesis inhibitor (such as puromycin) to the cells and compare the ability of these cells to insert new LDL receptors into the membrane compared with wild-type cells. If the protein synthesis of LDL receptors in the isolated cell strain is defective in any manner compared with the wild type, the former should be more sensitive to the effects of low doses of puromycin than wild-type cells would be.

40. You could microinject either of these tools and measure EGF-dependent Ras activity. If the activity of injected cells is decreased more than control, then EGF might be working through Ras.

8. Energy, Mitochondria, and Chloroplasts

1. True
2. False
3. False
4. True
5. False
6. False
7. False
8. True
9. True
10. True
11. False
12. True
13. False
14. True
15. False
16. obligate anaerobes
17. facultative anaerobes
18. lactic acid
19. chemiosmosis
20. proton
21. Hexokinase
22. pyruvate
23. Cristae
24. Oxygen
25. acetyl CoA
26. more
27. higher

28. 3
29. hydrolyze
30. synthase *or* synthetase
31. inner
32. NADH
33. 2; 1
34. CoQ *or* coenzyme Q (ubiquinone)
35. heme
36. peroxisomes
37. thylakoid; stroma
38. magnesium
39. water
40. higher
41. Planck's
42. carbon dioxide; oxygen
43. Phosphoenolpyruvate carboxylase
44. (c)
45. (e)
46. (b)
47. (e)
48. (a)
49. (c)
50. (b)
51. (e)
52. (c)
53. (d)
54. (e)
55. (e)
56. (a)
57. (c)
58. (a)
59. (c)

60. (b)

61. (e)

62. (a)

63. Sucrose was used to make the medium isosmotic with the chloroplast. If water, alone, had been used, the chloroplast would have swelled and ultimately ruptured owing to osmotic imbalance. Bacteria were preferentially located next to the chloroplast because they were probably either obligate or facultative aerobes. As such, they needed the oxygen being released by the chloroplast. Most bacteria of these types can sense a source of oxygen and migrate to it.

64. (a) Phenol red is a pH indicator dye that changes color when in the presence of protons. For instance, it turns purple in basic solutions and yellow in acidic conditions. Isolated thylakoid membranes can be stimulated with light so that protons are pumped to the inside of the disks. This creates a loss of protons from the medium outside the thylakoid membranes, and this can be reflected by a change in color (red to purple) of the phenol red, measured by a spectrophotometer.

(b) The wavelengths 680 nm and 700 nm are absorbed by the specialized chlorophyll molecules p_{680} and p_{700} present in PSII and PSI, respectively. The wavelength of 690 nm, however, is not absorbed as well. Therefore, the photosynthetic transport system is more active in the case of the two separate wavelengths.

(c) In cases where the intralumenal pH and the outside pH are identical, the proton-motive force is not sufficient to drive ATP synthesis. (The pH component of the proton motive force for chloroplasts is more important than the transmembrane potential.) When the intralumenal pH is 8.0 and the outside pH is 4.0, the protons go through the CF_0CF_1 complex in reverse (compared to *in vivo*). As a result, ATP is hydrolyzed to ADP + p_i.

(d) Thylakoid membranes are leakier to anions than are inner mitochondrial membranes. This is the underlying reason for the difference in the chloride diffusion data. The implication is that the proton-motive force of mitochondria relies more on the membrane potential differential; the proton-motive force of chloroplasts relies more on the pH differential.

65. The CF_0CF_1 complex is best observed using transmission electron microscopy and negative staining; PSI and PSII are best seen using freeze-fracture. The latter is possible because PSI and PSII are both integral membrane protein complexes. While CF_0CF_1 is also an integral membrane complex, it is best seen using negative staining.

66. The amount of ATP produced by chloroplasts is limited and ATP can't be stored in sufficient quantity to power the cells during periods of darkness. Furthermore, many plant cells (such as root cells) do not contain chloroplasts and thus need mitochondria to generate all of their ATP.

67. Isolate mitochondria from cells and then add radioactively tagged pyruvate. Compare the rate of entry of the tagged pyruvate into nontoxin-treated, control mitochondria with the rate of entry of the tagged pyruvate into toxin-treated mitochondria. A decrease in transport rate would suggest that the toxin might act as a pyruvate transport toxin.

68. Protons are, indeed, pumped to the intermembrane space but tend to leak to the cytosol through the outer mitochondrial membrane. They do so, in part, because they can diffuse easily through mitochondrial porin, a large protein in the outer mitochondrial membrane that can pass relatively large molecules (4000 to 5000 MW) and thus passes protons easily.

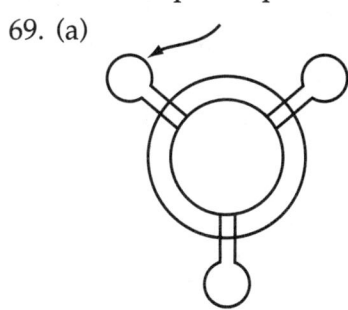

69. (a)

(b) H⁺ pumping

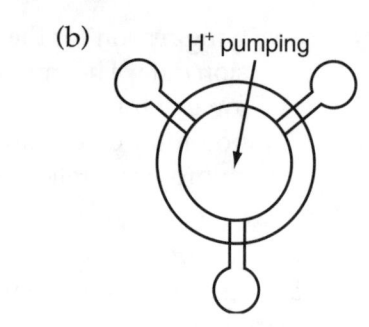

(c) Dinitrophenol is a weak, lipid-soluble acid that diffuses into the inside of the inside-out vesicle and transports protons from the inside of the vesicle, where they are in high concentration, to the outside of the vesicle, where they are in low concentration. This compromises the proton-motive force and therefore decreases ATP production.

(d) Valinomycin causes potassium ions to flow from the inside of the vesicle to the outside medium. (In the intact mitochondrion, potassium would flow from the outside of the mitochondrion to the inner matrix.) This flow occurs because potassium ions are moving from a positive potential (inner surface of the vesicle) to a negative potential (outer surface of he vesicle). This causes a decrease in the proton-motive force and therefore a decreased ATP production.

(e) Sodium azide is an electron transport inhibitor and therefore blocks the flow of electrons through the electron transport imtermediates. Electron flow is necessary for proton flow, so sodium azide decreases the proton-motive force. This leads to decreased oxidative phosphorylation.

(f) Oligomycin inhibits the F_0F_1 complex. It prevents oxidative phosphorylation because it prevents protons from flowing through the F_0F_1 complex. Oligomycin, however, does not directly affect the electron transport chain. Thus, the electron transport chain continues to operate, but the protons are unable to flow down their gradient. The proton gradient builds up over time, thus increasing the proton-motive force.

(g) Chloramphenicol is a mitochondrial protein synthesis inhibitor. Oxidative phosphorylation in inside-out submitochondrial vesicles does not depend on mitochondrial protein synthesis.

(h) Small amounts of detergents, even below their critical micelle concentration (concentration necessary to completely solubilize the membrane), causes the membrane to leak protons. This compromises the proton-motive force and, therefore, decreases oxidative phosphorylation.

(i) The intact F_0F_1 particle is capable of synthesizing ATP when present in intact, submitochondrial particles. Dissociated F_1, however, can no longer act as an ATP synthase, but, rather, has ATPase activity. Furthermore, this state of the F_0F_1 particle does not directly affect the operation of the electron transport system.

(j) The role of the F_0 particle is to conduct protons through the inner mitochondrial membrane.

70. The data with sodium azide, dinitrophenol and valinomycin suggest that rhodamine 123 retention is due to both the pH differential and the membrane potential differential generated by the proton-motive force. Rhodamine 123 retention is not, however, affected by protein synthesis inhibitors. The reason oligomycin increases rhodamine 123 retention is probably that this drug increases the proton-motive force by inhibiting the flow of protons through the F_0F_1 particle.

9. Cytoskeleton and Cell Movement

1. F
2. E
3. B
4. H
5. D
6. J
7. G
8. A

9. I
10. C
11. cytoskeleton
12. Actin
13. Myosin S1 (or The myosin head domain)
14. electron
15. Dystrophin
16. faster
17. **critical concentration**
18. myosin
19. optical
20. striation
21. sarcomere
22. myosin
23. actin
24. fibronectin
25. β
26. protofilaments
27. tubulin
28. MTOC or microtubule-organizing center
29. basal body
30. microtubules
31. (a)
32. (c)
33. (d)
34. (c)
35. (b)
36. (a)
37. (a)
38. (e)
39. (a)
40. (c)

Answers to Self-Testing Questions

41. (b)
42. (b)
43. (a)
44. (d)
45. (b)
46. (b)
47. (d)
48. (a)
49. (e)
50. (e)
51. (c)
52. (d)
53. (d)
54. (a)
55. (c)
56. False
57. True
58. False
59. True
60. True
61. True
62. False
63. True
64. False
65. True
66. True
67. False
68. True
69. False
70. True
71. True
72. True

73. True

74. False

75. True

76. AMP-PNP is a nonhydrolyzable analog of ATP. The fact that AMP-PNP caused binding of kinesin to the microtubule suggests that ATP, but not its hydrolysis, is important for the binding of kinesin to the microtubule. The lack of movement in the case of AMP-PNP, compared with movement when ATP is added, suggests that the hydrolysis of ATP is mandatory for this molecular motor to move its cargo on microtubules.

77. Kidney cells are polarized in the apical domain and typically feature microvilli that increase the absorptive area. Treatment by cytochalasin depolymerizes the actin filaments present in the microvilli, causing them to collapse into the cell. This process results in a decrease in cell surface area.

78. (a) Add $^{32}PO_4$ to cells going through mitosis. Immunoprecipitate lamin and separate it on an SDS gel. Localize phosphorylated lamin using autoradiography.

    (b) Purify nuclei from pre-mitotic cells, add γ labeled ATP and cytoplasm, immunoprecipitate lamin, and separate it on an SDS gel followed by autoradiography. This experiment, coupled with proper controls such as substituting cytoplasm with a salt solution, will verify that a factor is present in the cytoplasm that phosphorylates lamin. Now, fractionate the cytoplasm using a combination of centrifugation techniques (to remove organelles), chromatography, and gel filtration (to separate proteins). Test these fractions using the previous assay for lamin protein kinase activity. This purification scheme should isolate the lamin kinase.

    (c) Place a constant amount of lamin in solution in the presence of cytoplasm from cells in G1, S, G2, or M. Now, add γ [^{32}P]ATP to the mixture, incubate, and analyze as before with SDS gel electrophoresis and autoradiography. Finally, quantify the extent of phosphorylation of lamin by doing densitometry of the film, or alternatively by cutting out the corresponding bands from the gel and counting radioactivity using a scintillation counter or similar device.

79. (a) This treatment will remove most membranes, organelles, and proteins. Actin filaments and microtubules will also be removed if they are not highly cross-linked. The only remaining structures will be intermediate filaments. This could be verified by immunostaining the remaining filaments using a neuronal-specific intermediate filament antibody.

(b) Treat dorsal root ganglion cells with the microtubular inhibitors nocodazole or colchicine. If the axons retract, then microtubules are critical to the maintenance of the axonal morphology.

80. These data can be interpreted in two different ways. It is clear that disruption of actin filaments by cytochalasin causes an increase in the amount of intracellular laminin, as revealed by the increase in the radioactivity of the band in the lane cytochalasin D. One possibility is that laminin is retained (i.e., not secreted) by NHEK treated with cytochalasin D, and this is the reason underlying the increase in radioactivity. Alternatively, it's possible that the cytochalasin increases the synthesis of laminin without affecting the rate of laminin secretion. Thus, this experiment by itself does not verify if actin filaments regulate laminin secretion by NHEK cells *in vitro*.

81. Colchicine inhibits the polymerization of tubulin heterodimers into microtubules; whereas taxol promotes the polymerization of tubulin heterodimers into microtubules. Both can inhibit the mitotic spindle through different mechanisms and thus can affect cancer cells during the M phase of the cell cycle.

82. (1) Capping: microtubules can be capped at both ends by a variety of mechanisms. This stabilizes the tips of the microtubules so that there is little or no addition or removal of tubulin heterodimers. The best example of this process is the stabilization of the minus end of microtubules by the centrosomes or microtubule organizing centers.

(2) Critical concentration: the plus and minus ends have critical concentrations at which there is no net loss or addition of tubulin heterodimers. In both cases, and in contrast to capping described previously, there can be an exchange of tubulin heterodimers from both ends, but this is not mandatory.

(3) Treadmilling: In this case there is an addition of tubulin heterodimers at one end and removal of tubulin heterodimers at the other. Implicit in treadmilling is the notion that the rate of tubulin heterodimer addition is exactly equal to the rate of tubulin heterodimer loss. In this case there is a turnover of tubulin from one end (the addition end) to the other (the removal end).

83. When microtubules are assembled with GTP *in vitro* (and *in vivo*), GTP is hydrolyzed to GDP. Typically, the GDP-bound tubulin heterodimers in the microtubule are less stable than the GTP-containing tubulin heterodimers present as the "GTPcap" at the end of the microtubules. GMP-PNP cannot be hydrolyzed like GTP, and thus microtubules with tubulin heterodimers containing GMP-PNP are much more stable than microtubules with tubulin heterodimers containing GTP.

84. Telescoping occurs when the microtubule doublets are no longer held in register, one with the other. The protein nexin is responsible for this arrangement. Thus, it is possible that nexin was inadvertently compromised or removed in the preparation, resulting in telescoping.

85. Raw meat has a lot of actin. This actin binds the ingested phalloidin and so sequesters it, making it less toxic.

86. Add fluorescently tagged antibodies to living cells and do a fluorescence recovery after photobleaching experiment to demonstrate that the membrane protein does not move. Next, do the same with cells pretreated with cytochalasin. If protein X is associated with actin filaments, then the fluorescence recovery after photobleaching experiment might reveal enhanced mobility of protein X, indicating that it was directly or indirectly associated with actin filaments.

87. One approach is to do genetic studies. It might be possible to isolate a random mutant of this cell that does not demonstrate cytoplasmic streaming. If this is the case, one could purify the myosin and actin species present to determine if either of these molecules is defective. This study, however, would not be conclusive. Another approach would be to inject antibodies against actins or myosins into the cell. (You would need to ensure that these antibodies recognize the actin and myosin present in this novel organism.) If the injection of one of these antibodies (but not a control antibody) inhibits cytoplasmic streaming, then the molecule to which the antibody is specific can be implicated in the cytoplasmic streaming process.

88. Initiate F-actin polymerization. Once initiated, add myosin heads so as to decorate the actin filament. Next, wash away all excess, unbound myosin molecules. Now add excess (above the critical concentration) G-actin for a period of time necessary to achieve elongation of the filament. Fix for electron microscopy. You should see the decorated actin filament with thinner, nondecorated filaments extending from either end. The filament extensions should be shorter at the minus end than at the plus end, indicating that G-actin can add faster to the plus end.

10. Nervous and Immune Systems
1. E
2. B
3. G
4. C
5. H
6. I
7. J
8. A
9. F
10. D
11. True

Answers to Self-Testing Questions

12. False
13. False
14. True
15. False
16. False
17. False
18. True
19. True
20. False
21. True
22. True
23. True
24. False
25. True
26. True
27. False
28. False
29. False
30. True
31. False
32. True
33. False
34. True
35. True
36. False
37. False
38. True
39. True
40. True
41. axon
42. dendrites
43. retrograde; anterograde (orthograde)

Answers to Self-Testing Questions

44. action potential
45. closer
46. graded (or passive)
47. intracellular
48. skeletal; cardiac
49. active
50. NMDA
51. adenylate cyclase
52. neurotransmitter
53. close
54. rhodopsin
55. phosphodiesterase
56. Epitope
57. hapten
58. secretory component
59. B
60. instructive
61. clonal selection
62. Memory
63. thymus
64. tolerance
65. T
66. graft
67. intracellular; extracellular
68. macrophages
69. rough endoplasmic reticulum
70. bivalent
71. (d)
72. (e)
73. (b)
74. (a)
75. (d)

76. (c)
77. (b)
78. (d)
79. (d)
80. (d)
81. (e)
82. (a)
83. (b)
84. (d)
85. (b)
86. (e)
87. (b)
88. (a)
89. (a)
90. (e)
91. (c)
92. (b)
93. (e)
94. (d)
95. (a)
96. (e)
97. (c)
98. (e)
99. (b)
100. (a)
101. (c)
102. (b)
103. (a) The acetylcholine causes the muscle to contract, thus the on-cell technique would not be stable enough to do a recording.

　　(b) Add acetylcholine to the outside bath. If you detect current flow, then this would be an outside-out patch. If you don't, either this is an inside-out patch

Answers to Self-Testing Questions

or the preparation is a defective (i.e., nonresponsive) outside-out patch.

(c) You can't because it is an inappropriate technique to answer this question. You would have to do a receptor binding assay of some sort, possibly using isolated membranes.

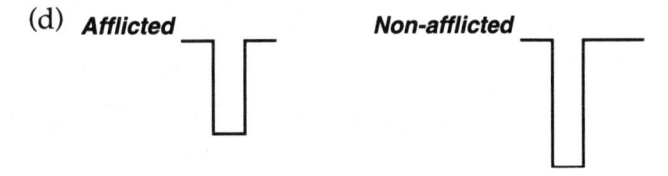

(d) **Afflicted**   **Non-afflicted**

(e) Atropine extends the life of acetylcholine in the synaptic cleft. This, in turn, would cause more channels to open.

(f) Lysine is positively charged and, as such, might restrict the passage of sodium and potassium through the channel more than the glutamic acid residues.

(g) This is a difficult question to answer. If genes coding for the correct subunits could be inserted into afflicted cells, new receptors encoding the proper M2 helix might be constructed and the disease might be partially relieved. The problem, of course, is how one might "insert" these genes into all the skeletal muscle cells of an afflicted individual. Thus in theory this sounds interesting, but in practice there are some major hurdles that need to be overcome.

104. Any cell needs to expend energy to maintain osmotic balance. Part of this maintenance is the constant pumping of sodium out of and potassium into the cell. This requires ATP. The greater the surface to volume ratio, the harder these pumps need to work in order to maintain the proper chemical equilibria.

105. You could use the radioisotope ^{131}I-α-bungarotoxin as a tool to analyze receptors. Add this label to postsynaptic cells with de-localized receptors, film the preparation for autoradiography, and determine the localization of the receptors.

106. (a) Immunocytochemistry should demonstrate "particles" in cells that are equivalent to vesicles. (Vesicles as such cannot be discerned easily using fluorescence microscopy). Alternatively, if the neuropeptide were soluble in the cytoplasm there would be a diffuse, homogeneous pattern of fluorescence signal originating from the cell.

(b) Do immunocytochemistry or standard transmission electron microscopy to determine if nocodazole disrupts microtubules.

(c) Cytochalasin B inhibits the formation of actin filaments that are at the leading edge of a growth cone and synaptic processes.

(d) An increase in intracellular calcium in the axonal terminal due to opening of plasma membrane Ca^{2+} channels is necessary for both neurotransmitter vesicle and neuropeptide vesicle fusion with the presynaptic membrane.

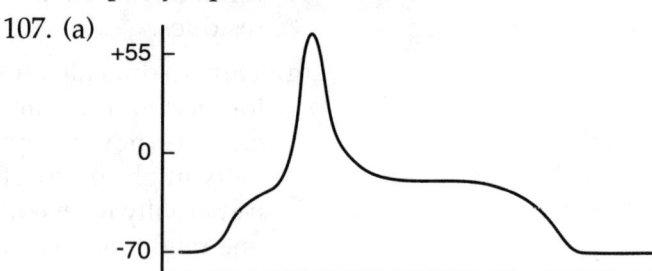

107. (a)

+55

0

-70

(b) Decreased extracellular calcium should diminish delayed repolarization if voltage-gated calcium channels are opening. An alternative would be to add a toxin specific to voltage gated calcium channels, but this is more complicated because each type of voltage-gated calcium channels is sensitive to different sets of toxins.

(c) Activation of second messengers by some postsynaptic receptors will result in the phosphorylation of channels which, in turn, changes their conductance profiles.

108. (a) It is in approximate osmotic balance since the number of milliosmoles inside = 470, whereas the

number of milliosmoles outside = 452.

(b) Ew = +58mv,

Ex = +81mv

Ey = −45mv

Ez = +23mv

(c) Use the Goldman equation here as in the following manner.

$$V = 58 \log_{10} \frac{0.2(250) + 1.0(50)}{0.2(10) + 1.0(300)} = -29 \text{ mv}$$

109. (a) A + C are additive and summed together exceed the threshold for this neuron.

(b) Possibly A and C are excitatory and result in a depolarization of the postsynaptic membrane, whereas B is inhibitory and results in a hyperpolarization of the postsynaptic membrane. When all three are stimulated together the inhibitory influence of B is enough so that the stimulatory influence of A + C is below threshold. Hence, an action potential fails to materialize.

(c) They are at the axon hillock and the nodes only.

110. The monoclonal antibody is specific for one epitope that is unique for this novel protease. The polyclonal antibodies could be labeling more than one protein for at least two different reasons. First, one of the polyclonal antibodies might be labeling an epitope that is a domain shared by other proteases within the same family. Hence, this type of antibody is labeling the same epitope in several proteins. Alternatively and more probably, the polyclonal antibodies recognize different epitopes and thus are cross-reacting with other proteins in the preparation.

111. The graft from a different individual would be much more likely to cause graft rejection. This is due to the major histocompatability antigens present on the grafted tissue that are recognized as foreign by the immune system. Heart, liver, pancreas, and lung transplants are currently all grafts from others and thus immunosuppressant drugs need to be administered.

112. At high concentration there is excess antigen so that antibodies can't cross-link with other antibodies associated with antigen. At lower concentration antibodies can now cross-link with each other as shown below.

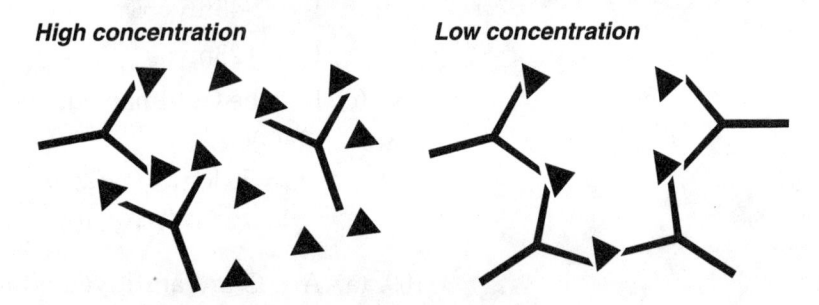

High concentration   **Low concentration**

113. (a) Lymphocytes migrate from the liver and bone marrow and take up residency in the thymus where they further mature. Then they leave the thymus and move to peripheral lymph tissues such as lymph nodes.

(b) Lymphocytes have receptors that bind to the endothelial cells that facilitate migration into the thymus gland.

(c) One approach would be to pretreat the lymphocytes or the endothelial cells with a protcase. This enzyme should cleave the receptors and, as such, binding of the two cell types might be prevented.

114. There are many possibilities for solving this problem. One approach is to do a Western blot of proteolytically digested antigen. If there is only one epitope, a single band would appear on the blot. If there is more than one epitope in Wop, and assuming the different proteolytic fragments are of different size and can be distinguished from each other, multiple bands would appear on the blot.

115. (a) The second injection was probably given between week 5 and week 6.

(b) The large increase in circulating antibodies is due to the second injection probably given between weeks 5 and 6. The primary injection generated memory cells that were subsequently activated by the second immune response from the second injection.

(c) The antimyelin antibodies generated by the rabbit may have reacted against the rabbit's own myelin. If this were the case, the antibodies might be partly responsible for an autoimmune response that results in a demyelination, similar to what occurs in multiple sclerosis.

116. (a) One approach would be to use a panning (i.e., selective surface) technique utilizing the fact that T cells, but not other blood cells, have a Thy-1 marker on its surface. Thus an antibody to Thy-1 could be applied to the bottom of the cell culture ("pan") plate and used to retain T cells while others are washed off.

(b) Cytotoxic T cells kill cells through a mechanism of apoptosis or programmed cell death. One of the classic characteristics of apoptosis is clumped chromatin in the nucleus.

(c) This "conditioned medium" experiment suggests that the cytotoxic T cells responded to the epithelial cells by secreting toxins into the medium that initiated apoptosis.

(d) Cytotoxic T cells need to associate with the major histocompatablity antigen containing the peptide virus to initiate cell death. The anti-major histocompatability antibodies prevent this association.

(e) This suggests that T cells recognize both self and foreign determinants together but cannot distinguish either separately.

11. Extracellular Matrix and Cell-Cell Interactions

1. E
2. B
3. G
4. I
5. A
6. D

Answers to Self-Testing Questions

7. J
8. F
9. C
10. H
11. False
12. True
13. True
14. False
15. False
16. True
17. False
18. True
19. True
20. False
21. True
22. False
23. True
24. False
25. True
26. False
27. False
28. True
29. False
30. False
31. (c)
32. (c)
33. (a)
34. (b)
35. (a)
36. (e)
37. (c)
38. (b)

Answers to Self-Testing Questions

39. (b)
40. (d)
41. (a)
42. (e)
43. (d)
44. (b)
45. (e)
46. (b)
47. (d)
48. (a)
49. (d)
50. (b)
51. (e)
52. (e)
53. (d)
54. (b)
55. (b)
56. (a) Lysyl oxidase is an enzyme that forms an aldol cross link between two lysine residues in collagen. This extracellular enzyme thus links two neighboring collagen helices together. A defect in this enzyme would lead to aberrant collagen fibril formation.

    (b) Vitamin C, or ascorbic acid, is critical to the proper hydroxylation of the collagen molecule in the rough endoplasmic reticulum. It is a cofactor for prolyl 4-hydroxylase and prolyl 3-hydroxylase that catalyzes the hydroxylation. Without vitamin C, the collagen triple helix would not form properly and be degraded in the rough endoplasmic reticulum.

    (c) These propeptides are cleaved in the extracellular domain prior to the assembly of the collagen fibril. They help direct the final assembly of the collagen triple helices into collagen fibrils. Defective propeptides, therefore, might lead to abnormal collagen fibrils.

57. One can infer from these observations that the collagen fibers in the cornea and skin are stable and do not undergo a turnover (cycles of proteolysis and synthesis), whereas the collagen surrounding blood vessels does undergo such a turnover.

58. The epithelial cells could either be associating with each other through an X-cadherin/X-cadherin homophilic interaction or an X-cadherin/other heterophilic interaction. One method to investigate this idea is to transfect the gene encoding X-cadherin into the neuronal cells and now determine if the neuronal cells adhere to the epithelial cells. If they do, then X-cadherin is working in a homophilic manner. If not, the X-cadherin is working in a heterophilic manner.

59. Place fibronectin on the bottom of a culture plate and let dry. Seed epithelial cells on the surface to ensure that they adhere. Now repeat this procedure again but, prior to seeding the epithelial cells on the plate, bathe the epithelial cells in a solution containing RGD. If the cells don't adhere to the bottom of the cell culture dish, it is possible that the RGD peptide in the bath bound to the integrin receptors making them unable to bind to fibronectin coated on the bottom of the cell culture plate.

60. Neonatal N-CAM has much more sialic acid attached to it then does adult N-CAM. This is revealed in the SDS-PAGE Western blots with the Neonatal N-CAM having a larger, apparent molecular weight than adult N-CAM. This extra sialic acid keeps the neonatal neurons apart from each other so that they can slide past one another during development. In the adult, however, it is critical that neurons adhere tightly to each other. Otherwise, synaptic communication will be interrupted.

61. Collagen helices, like DNA helices, denature in response to heat. In both cases hydrogen bonds are broken and this change in helical arrangement can be monitored using a spectrophotometer. By measuring conventional collagen to the new type of collagen in this manner you might note a difference in their abilities to withstand heat. If the new collagen had fewer hydrogen bonds, the data would appear similar to the following:

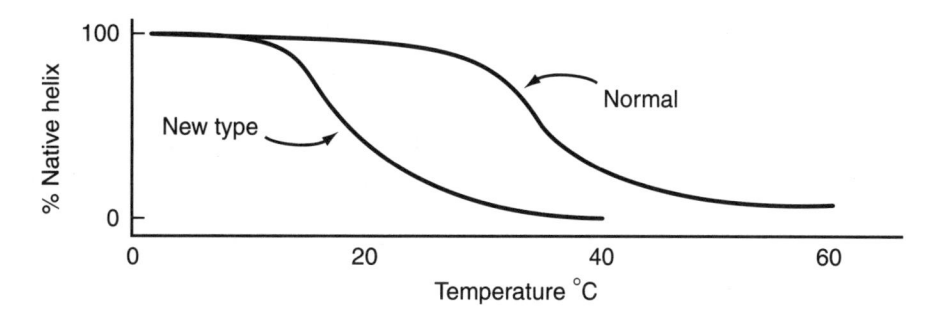

62. (a) Many cell adhesion molecules, such as E-cadherin present in keratinocytes, need calcium to function. If higher concentrations of calcium were included in the medium, the cells would clump together and not develop into a monolayer.

(b) The proteoglycans might be important in this regard for three reasons. First, it would add a hydration factor that would enhance the hydration of the Type I collagen gel. In addition, proteoglycans are known for binding growth factors, such as Fibroblast Growth Factor. By so doing, they position the growth factors near the cells and immobilize them, thus allowing the factors to have a continuous effect on their receptors. Furthermore, the proteoglycans might act as a storage device, constantly releasing hormone to the keratinocytes.

(c) One approach is to determine if this laminar structure has all the components characteristic of a basal lamina. One method of demonstrating this would be to stain for Type IV collagen and laminin—two key components of the basal lamina. This could be accomplished using the appropriate antibodies and ultrastructural immunocytochemistry. (Antibodies that are specific for Type IV collagen but not Type I collagen are available.)

(d) You could label the preparation with ^{3}H-proline. Proline is a common amino acid in collagen, but not as common in other proteins. The newly synthesized keratinocyte collagen, but not the Type 1 collagen in the gel, would incorporate the ^{3}H-proline and then be deposited as Type IV collagen in the basal lamina.

(e) Human epidermis contains numerous desmosomes, **hemi-demsosomes** and keratin filaments. Thus the electron microscope could be used to detect the presence of these structures. Western blots, on the other hand, would be useful to determine if skin-specific keratins are present in the *in vitro* epidermis.

(f) The *in vitro* epidermis is twice as leaky as the *in vivo* epidermis.

(g) (i) The cancer cells are probably secreting collagenase, an enzyme that can hydrolyze collagen. One method that could be used is to add an inhibitor to collagenase to determine if this process is inhibited. Alternatively, adding collagenase antibodies to the extracellular medium might bind this extracellular enzyme so that it is not as effective, thus inhibiting the cell migration process.

(ii) Fibronectin is a linker molecule that promotes adhesion of cells to extracellular matrices. Cancer cells, by nature, are very mobile. Synthesis and secretion of fibronectin would make these cells less mobile and less able to migrate through the porous membrane.

63. (a) Lightly treating the cells with a protease should make them adhere less to each other if cell adhesion molecules are mediating part of this aggregation.

(b) One possible approach would be to grow the cells in a large volume of medium and compare the ability of these cells to aggregate when the same number of cells are grown in a smaller volume of medium. In the former case the extracellular linker molecule might be less concentrated and thus many of the cell adhesion molecules would not be linked. In the latter case there would be a greater concentration of the extracellular linker molecule, and more of the cell adhesion molecules would be linked. Thus, these cells would adhere to each other more tightly.

12. The Cell Cycle and DNA Repair

1. B
2. C
3. E
4. A
5. D
6. True
7. True
8. False
9. True
10. False
11. True
12. False
13. False
14. False
15. True
16. True
17. False
18. True
19. False
20. True
21. metaphase
22. G_2
23. **oocytes**
24. box
25. MPF
26. lamins
27. karyomere
28. dephosphorylation
29. myosin (the light chain)
30. MPF
31. cyclin

32. Size
33. START
34. **Nondisjunction**
35. (a)
36. (e)
37. (d)
38. (b)
39. (a)
40. (a)
41. (b)
42. (e)
43. (b)
44. (c)
45. (b)
46. (b)
47. (c)
48. (b)
49. (c)
50. First, these egg cells are large and thus easy to microinject with proteins that regulate the cell cycle. Second, they are plentiful and easy to harvest. Third, they divide synchronously when stimulated with sperm cells. Thus, biochemical activities that occur at certain points in the cell cycle can be easily analyzed.
51. First, temperature-sensitive and cold-sensitive mutants have been made that stall at different parts of the cell cycle owing to enzymatic deficiencies that can be subsequently elucidated. Second, genes providing this missing enzyme can be identified through genetic analysis. If mutant yeast are stalled in the cell cycle at the nonpermissive temperature, the missing enzyme provided by the proper transfected gene will permit the yeast to move through the cell cycle. Third, they replicate quickly in culture and are easy to grow. Fourth, they can be in either a haploid or a diploid state, making genetic analysis easier than in other diploid eukaryotic cells.

Answers to Self-Testing Questions

52. (a) Effect—DNA synthesis in the G_1 nucleus is initiated. This is investigated by using ^{3}H-thymidine. The implication is that there is a trans-acting factor that can initiate DNA synthesis.

    (b) Effect—DNA synthesis in the G_2 nucleus is not initiated. This is investigated by using ^{3}H-thymidine. This suggests that the G_2 nucleus has some type of block or barrier that prevents erroneous duplication of the DNA subsequent to the S phase.

    (c) Effect—The G_1 nucleus breaks down. This is investigated by transmission electron microscopy. The implication is that there is a trans-acting factor in the mitotic cell that causes the dissolution of the nuclear membrane.

    (d) Same effect as part (c).

53. **G_1**—The first step is to synchronize the cells in G_0. This could be done in a number of ways, including amino acid deprivation or contact inhibition. Next, the cells are given ^{3}H-thymidine and released from their stalled position in G_0. The time necessary for the first cells to reach S, as evidenced by the incorporation of the ^{3}H-thymidine, is an approximation of the G_1 compartment time.

    S—The first step is to determine the total cell cycle time, i.e., the time necessary for actively cycling cells to go from n to $2n$ cells. Next, ^{3}H-thymidine is given to actively asynchronously dividing cells, and the fraction of cells that incorporate the ^{3}H-thymidine after a 30-minute incubation time is determined. Multiplying this fraction by the total cell cycle time yields an approximation of S compartment time.

    G_2—^{3}H-thymidine is given to actively asynchronously dividing cells, and the time necessary for the first appearance of ^{3}H-thymidine–labeled mitotic figures is determined. This is a rough approximation of G_2 time, since cells leaving S will incorporate the ^{3}H-thymidine and should be the first to go into mitosis.

    M—The first step is to determine the total cell cycle time. Next, the fraction of cells with mitotic figures (i.e., mitosing cells) is calculated by simple microscopic inspection. No ^{3}H-thymidine is necessary for this step. This fraction, multiplied by the total cell cycle time, yields the time spent in M.

54. (a) Lane A shows purified cyclin, but lanes B–D show cyclin as a larger molecular-weight complex. The reason for this difference is because cyclin B is only one part of the MPF heterodimer. The other part is the protein kinase cdc2 (cdk1). The protein kinase is not present in lane A, but is present in lanes B, C, and D. Also note that there is much less cyclin noted in lane E, and it has a smaller molecular weight than seen in the previous lanes. This is due to the fact that cyclin B degrades at the onset of anaphase. Thus, the antibody might be binding to smaller hydrolytic fragments of cyclin B, hence the smaller molecular weight.

(b) Cell mutants containing nondegradable cyclin B have been made. These experiments have shown that when cyclin B cannot degrade, the cells can never leave mitosis, i.e., they go into mitotic arrest.

(c) You could immunoprecipitate MPF using the cyclin B antibody and then assay for protein kinase activity using histone H1 as a substrate.

(d) You could remove the nucleus from the egg and assay for cyclin B during subsequent divisions of the embryo. (Fertilized eggs will go through the first few divisions without a nucleus because the mRNAs necessary are already stored in the cytoplasm.)

(e) You could add a protein synthesis inhibitor, such as cycloheximide, to see if the increase in the Western blot cyclin B staining seen in part (d) is blocked to any extent.

(f) Cyclin B would not degrade in late anaphase and, so, would appear similar to that in prophase and metaphase (lanes C and D).

Answers to Self-Testing Questions   281

(g) Ubiquitin marks cyclin B for destruction. If ubiquitin can't attach to cyclin B, then cyclin B won't degrade into cyclin peptides. Thus, the cyclin B staining pattern during late anaphase would be similar to that in prophase and metaphase.

(h) Colchicine disrupts microtubules in mitosis. By so doing, a checkpoint control is activated that senses that the microtubules are not properly aligned. Thus, the cells stall at metaphase. MPF activity is high at this time and doesn't fall until late anaphase. Thus, MPF activity does not decrease, because the cells can't proceed to anaphase in the presence of colchicine.

55. (a) Purified MPF could be added to purified lamins A, B, or C in the presence of γ AT^{32}P. Next, these proteins could be analyzed for phosphorylation using gel electrophoresis and autoradiography.

(b) Site-directed mutants for lamins A, B, or C could be produced and transfected into recipient cells. These mutants should have serine, which is normally phosphorylated by MPF, converted to alanine, which can't be similarly phosphorylated. Conventional light microscopy could be used to examine the ability of daughter nuclei to separate in anaphase. In experiments of this type, cells with mutant lamin A fail to proceed through mitosis properly.

56. The cyclin D-Cdk2 protein kinase activity mirrors the mRNA levels. This is because the mRNA for cyclin D, as well as most cyclins, is very unstable. This ensures that the cyclin D level drops as soon as the cells pass the G_1 restriction point. The answer therefore appears as shown at the top of page 282.

Northern blots of cyclin D mRNA

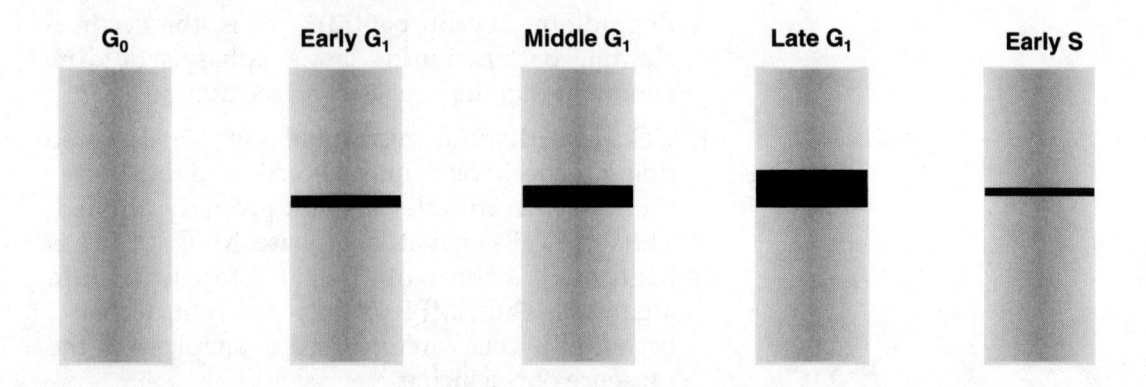

G₀ | Early G₁ | Middle G₁ | Late G₁ | Early S

Relative levels of cyclin D-Cdk2 protein kinase activity

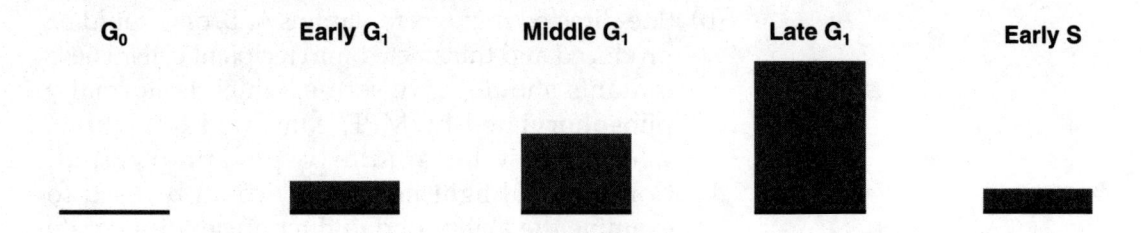

G₀ | Early G₁ | Middle G₁ | Late G₁ | Early S

13. Cancer and Cell Death

1. Metastatic
2. carcinoma
3. Melanoma
4. promoters
5. papilloma
6. collagenase
7. multidrug resistance
8. oncogene
9. tumor suppressor
10. *src*
11. **retrovirus**

Answers to Self-Testing Questions

12. transformed
13. apoptosis
14. apoptosis
15. apoptotic
16. (c)
17. (a)
18. (e)
19. (c)
20. (b)
21. (c)
22. (d)
23. (c)
24. (a)
25. (e)
26. (a)
27. (e)
28. (a)
29. True
30. True
31. True
32. False
33. False
34. True
35. False
36. True
37. False
38. True
39. True
40. True
41. False
42. False
43. False

44. Graph A shows cells entering "crisis," a condition in which they ultimately die. This can be realized from the "death curve" starting at day 6. Graph B shows cancer cells. Note that, as they continue to divide, they reach the time when they should reach confluency if they were normal (see graph C). Graph C shows normal growth of noncancer cells. The cells divide initially at subconfluent densities (days 1–3) and then reach "confluence," where they are contact-inhibited, i.e., stop dividing. This results in a plateau. Graph D depicts normal cells that reach confluence (contact inhibition); however, a spontaneous mutant probably arises that produces a cancer cell, so at day 9 cell division resumes.

45. These data imply that the carcinogen is an indirect-acting carcinogen. Note that the putative carcinogen was unable to cause cancer directly (condition B), but could do so when in the presence of the liver cells (condition C). The cytochrome P_{450} of the liver cells probably converted the putative carcinogen into an active (i.e., electrophile-containing) carcinogen. This would explain the log growth characteristics in later days that is not seen in cultures A and B.

46. Graph A represents stationary growth of sympathetic neurons in NGF-supplemented media. Graph B demonstrates that NGF is necessary for survival, as stated in the question. Note that in graph C the protein synthesis inhibitor is able to extend the life expectancy of sympathetic neurons. This is because NGF withdrawal probably causes "suicide genes" (apopotic genes) to be turned on. This can be inferred because the protein synthesis inhibitor probably blocks the synthesis of the protein products of these suicide genes. There is no implication for necrosis revealed in these experiments.

47. There are a few methods that could be used. First, in rare cases some transformed cells will express their oncogenic phenotype in one type of extracellular matrix but express their normal phenotype in a different extracellular matrix. Alternatively, temperature-sensitive mutants exist that exhibit their oncogenic phenotype when grown at 39°C, but exhibit their normal phenotype when maintained at 37°C.

Answers to Self-Testing Questions 285

48. In graph A, tumor suppressor genes are working to suppress the oncogenic phenotype. Hybridomas, especially cross-species hybridomas such as this one, tend to lose genes and chromosomes over time. It is possible that the hybridoma loses the gene coding for a tumor suppressor gene, resulting in the cancer phenotype depicted in graph B. This type of analysis is often useful in identifying what chromosome contains the tumor supppressor gene.

49. Myristic acid helps the oncoprotein attach to the plasma membrane. Without the myristic acid, the src oncoprotein floats unattached in the cytoplasm. While it can still act as a protein kinase (data not in question), it is unable to effect oncogenesis because of its cytoplasmic location.

50. Treatment of cells with the phorbol ester activates the protein kinase C system. In some cells, such as the ones in this problem, the presentation of phorbol esters causes the cells to divide continuously in the presence of this carcinogen. One could regard this performance by these cells as "phenotypic" of cancer because the cell division activity appears to be cancer-like, but there has been no change in the genes. This is substantiated because removal of the phorbol ester results in cell cycle arrest (the plateau in A). In B the cells are treated by a carcinogen that probably acts as an initiator. Initiators are essential in order for phorbol esters to cause a genotypic activation of cells to the cancer phenotype. Phorbol esters in this case act as promoters.

51. This is a peculiar cell. Note that it has normal characteristics when grown in standard cell culture, but when grown in soft agar (graph B), it divides as if it were a cancer cell. The soft agar experiment is very diagnostic because most normal, substrate-dependent cells will not grow in soft agar. Thus, this cell does have oncogenic characteristics when maintained in this environment. Yet, the data in graph A suggest that it does not form multilayers—a typical characteristic of cancer cells. Possibly there is some complicated interaction between the environment that acts as a permissive condition for the cell to exhibit its cancer phenotype.

52. There are many possible approaches to this problem. Apoptotic (programmed for cell death) and necrotic cells are different from each other in many ways. For instance, necrotic cells have leaky membranes, but this is not true with apoptotic cells. Thus, an assay could be developed that assesses enzyme leakage in both cases. But the hallmark of apoptotic cells is a DNA ladder. Thus, the genomic DNA from these dying skin cells could be harvested and separated on agarose gels. If the cells died through necrosis, a "smear" would be seen, whereas if the cells died through apoptosis, a DNA "ladder" would be revealed.

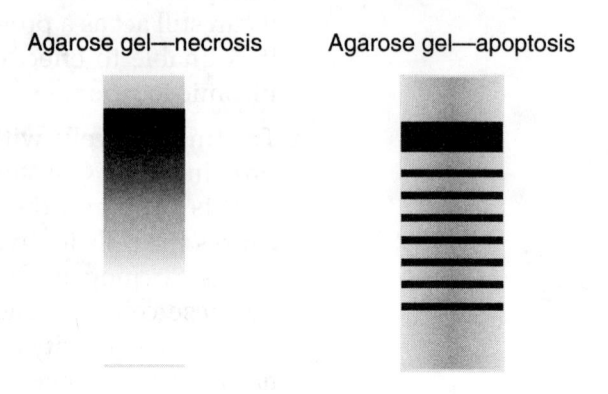

Agarose gel—necrosis   Agarose gel—apoptosis

14. Development and Differentiation

1. yolk
2. vegetal; animal
3. gray crescent
4. blastomeres
5. fate
6. Spemann's; blastopore
7. notochord
8. neural
9. somites
10. Induction
11. Spemann's organizer
12. zona pellucida

Answers to Self-Testing Questions

13. chimera
14. totipotent
15. Lineage
16. founder
17. anchor
18. heterochronic
19. apoptosis
20. syncytium
21. morphogen
22. germ
23. (a)
24. (d)
25. (a)
26. (b)
27. (d)
28. (a)
29. (c)
30. (d)
31. (b)
32. (d)
33. (c)
34. (e)
35. (c)
36. (e)
37. (b)
38. (a)
39. (e)
40. (b)
41. (a)
42. (d)
43. (b)

44. (a)
45. True
46. False
47. False
48. False
49. True
50. True
51. True
52. False
53. True
54. False
55. False
56. True
57. True
58. False
59. True
60. False
61. False
62. Cell division in most animal cells is highly dependent on protein synthesis. Transcription inhibitors usually block the cell cycle by inhibiting the synthesis of crucial proteins needed during the G_1 phase of the cell cycle. Blastomeres, however, use the mRNA and proteins stored in the egg yolk for cell division. This means that the blastomeres can have very fast (30-min.) cell cycles, because the cells are less dependent on transcription and translation.
63. There are at least two possible approaches. First, you could add specific inhibitors that block the movement of these ions so that their accumulation might be inhibited. You could then determine if the inhibitors blocked the formation of the blastocoel. Or, you could grow embryos in the presence of the radioactive isotopes of the ions and find which, if any, accumulate in the blastocoel. This second approach, however, is not sufficient by itself to show a causal relationship.

64. Calcium chelators decrease the availability of free calcium in solutions. The ability of cells in amphibian and other embryos to recognize position and to undergo morphogenetic movements is partly under control of the calcium-dependent cell adhesion molecules such as cadherins. Cadherins need to bind calcium in order to function. Thus, EGTA or EDTA might prevent the proper functioning of cadherins, which, in turn, would prevent proper development of the embryo.

65. The easiest approach to this experiment is to microinject anti-integrin antibodies into the developing embryo. Compare the subsequent development with that of control antibody-injected embryos. If the embryo injected with anti-integrin antibodies reveals defective embryogenesis, then integrins might be involved. A similar experiment using fibronectin antibodies could also be used to see if a fibronectin-integrin link might be involved.

66. There are at least two approaches that can be taken. First, you could stain neural crest cells with a vital dye and track them as they migrate. This method has several disadvantages. First, the dye might kill or adversely affect the migrating cell. Second, the dye might be lost. Third, the dye might be diluted if the cells divide. Nonetheless, this technique has been successfully used to accomplish this task. The second possibility is to transplant neural crest cells from a related, but different, embryo into the neural crest area. These transplanted cells need to be distinguishable from the cells in the recipient embryo so that they can be tracked during migration. Quail embryo cells can be transplanted to chick embryos and tracked because the quail cells have a distinctively staining heterochromatin not present in chick cells.

67. The amphibian embryo develops by both regulative development and localized determinants. In regulative development, cell to cell interaction is all important. Localized determinants, however, relate to the asymmetric distribution of molecules (mRNA in particular) in the embryo that influences body axis and organ development. Mechanical disruption of the egg

can disturb the distribution of mRNA and thus disturb the body plan of the future embryo.

68. One could transplant cells from one part of a donor embryo (e.g., position B) to a different part of a recipient embryo (e.g., position A). If the B cells become A cells, then determination hasn't yet occurred. If the B cells remain as B cells, then determination has most probably occurred.

69. *In situ* hybridization can be used to detect the expression of specific mRNAs in individual cells through the binding of a complementary probe to these mRNAs. They can show, for example, that the MYOD muscle genes are expressed primarily in somites.

70. To confirm cytoplasmic memory, a nucleus could be injected into the anucleate cytoplasm of another cell that is suspected of having undergone cytoplasmic memory. If the gene expression of the injected nucleus parallels that of the removed nucleus (possibly as revealed by *in situ* hybridization), then cytoplasmic memory can be implicated. To confirm paracrine memory, cells suspected of undergoing paracrine memory could be transplanted to a different area in another embryo. If the neighboring cells in the recipient embryo are induced to differentiate to the same type as the transplanted cell, then a paracrine memory mechanism can be implicated.

71. Lasers can be focused on single cells, and the cell can be killed. If the anchor cell is terminated in this manner, a vulva will not develop from underlying hypodermal cells.

15. Gene Expression in Eukaryotes and Prokaryotes

1. (a) X-gal acts as a substrate and therefore as a marker for the enzyme β-galactosidase. X-gal is converted into a blue pigment by β-galactosidase. Thus, all the bacterial colonies that produce the enzyme would appear blue.

    (b) The color of wild-type *E. coli* colonies would be blue when IPTG is added to the medium because, when IPTG induces the *lac* operon, β-galactosidase is produced in high amounts.

(c & d) *Cis*-acting. Operator-mutant. Operator cannot bind the *lac* suppressor.

Trans-acting. *lac I* mutant. *lac* suppressor cannot bind to the operator.

(e) Class A—The additional mutation may be on the *lac* suppressor. This mutation may have caused a structural conformation change in the suppressor molecule that enables binding to the mutant operator. These new mutants exhibit normal wild-type regulation of the *lac* operon.

Class B—These may have the mutation on the β-galactosidase gene, leading to an inactive enzyme. These mutants have a white phenotype even when IPTG is added.

(f) Class A mutants have a mutant *lac* suppressor that is unable to bind to the transfected *lac* operator. This results in the constitutive expression of β-galactosidase from the transfected construct. The colonies will be blue regardless of the presence of IPTG.

Class B mutants have wild-type *lac* suppressors and a defective β-galactosidase gene. The normal suppressor is able to bind to the operator on the transfected construct and normal wild-type regulation of β-galactosidase expression will occur. The colonies will be white when no lactose or IPTG is present. They will be blue when lactose or IPTG is present.

2. (a) A TATA box around −30 to −25 and initiator elements that span the transcription start site are generally found upstream of highly expressed genes such as globin, immunoglobulins, and ovalbumin.

(b) High amounts of G and C in the upstream region may indicate the presence of a promoter.

(c)

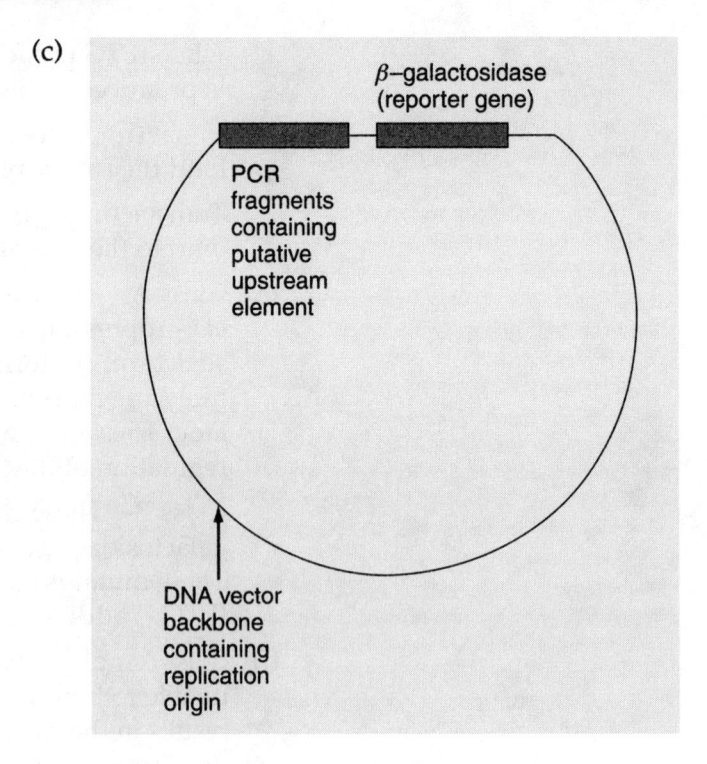

[Diagram: Circular plasmid showing PCR fragments containing putative upstream element adjacent to β-galactosidase (reporter gene), with DNA vector backbone containing replication origin.]

(d) If the upstream segment attached to the β-galactosidase gene happens to contain an upstream promoter or enhancer element, it will drive expression of β-galactosidase higher, and will be detected in the quantification assay.

(e) X-gal forms a blue-colored product when β-galactosidase acts upon it. This product can be quantitated by using a colorimeter.

(f) X is a mammalian gene. The upstream promoter or enhancer element may function optimally only in a mammalian cell, as transcription initiation in eukaryotes is a much more complicated process than in prokaryotes.

3. False. This mutant will not be able to be induced by lactose because lactose does not bind to the regulator protein.

4. True

5. False. The operator binds the regulator (*lac I*) protein.

6. True
7. True
8. True
9. True
10. True
11. True
12. False. At the nucleolus, the synthesis of ribosomal RNA occurs.
13. polycistronic
14. TATA box
15. histones; arginine; lysine
16. heterochromatin; euchromatin
17. ρ
18. removal of an intron; modification of certain bases
19. ferritin; ALA synthase
20. female
21. **lytic; lysogenic**
22. cI (lambda repressor)

glossary

α-Amino Group The amino group (–NH$_2$) attached to the α carbon of an amino acid.

α-Carboxyl Group The carboxyl group attached to the α carbon of an amino acid.

α-Helix A form of secondary structure of proteins in which the linear sequence of amino acids is folded into a right-handed spiral. This spiral structure is stabilized by hydrogen bonds formed between the oxygen of each peptide bond and the three amide residues farther in the sequence.

Abbe's Equation A mathematical equation developed by Abbe, giving resolution in both light and electron microscope systems. The equation is $D = 0.61 \lambda / N \sin \alpha$, where D = the distance between two just barely observable objects, λ is the wavelength of light, and N is the refractive index of the medium.

Acetyl Coenzyme A (Acetyl CoA) A small water-soluble molecule that is linked to coenzyme A. It is generated through pyruvate from glycolysis and through fatty acids and amino acids. It is the major metabolite of the Krebs cycle. It transfers its two-carbon acetyl group to citrate at the beginning of the cycle.

Acetylcholine A neurotransmitter secreted by cholinergic neurons. It is stored in small, electron-clear vesicles and released during action potentials. It interacts with its receptors and can activate second-messenger systems or, as in the case with muscle, open a sodium-potassium channel, causing a change in membrane potential in the postsynaptic cell.

Actin One of the most abundant proteins in eukaryotic cells. It is responsible for structure and cell movement. Half of actin is in globular form, called G actin; the other half is in filamentous form, called F actin. F actin has a polarity, with (+) and (–) ends.

Action Potential A change in membrane potential that occurs in excitable nerve cells. It lasts for 2 milliseconds and is unidirectional. It is caused by the depolarization of the cell membrane and the subsequent activation of voltage-sensitive sodium channels. The action potential can be blocked by neurotoxins such as tetrodotoxin (TTX) from the puffer fish.

Activated tRNA Enzymes known as aminoacyl-tRNA synthetases link amino acids to cognate tRNA molecules at the 3' terminus (of tRNA), using ATP. The resulting aminoacyl-tRNA molecules retain the energy of ATP and are known as activated tRNAs.

Active Transport A process by which a molecule can be transported across a membrane, usually against its concentration gradient. This process necessitates energy input, usually in the form of ATP. The sodium-potassium pump (ATPase) is an example.

Adherens Junction A specialized cell junction often found in epithelial cells and fibroblasts. The membrane of one cell is directly attached to the membrane of another. Actin filaments are linked to the junction. Focal contacts on fibroblasts are examples of this type of junction.

Aerobic Describing a process that requires oxygen, such as the production of ATP in mitochondria. Alternatively, it can refer to an organism that uses oxygen for metabolic processes, such as an aerobic bacterium.

ALA Synthase Aminolevulinic acid synthase is the key regulatory enzyme of the heme biosynthetic pathway. It catalyzes the synthesis of aminolevulinic acid from the substrates succinyl CoA and glycine.

Aldehydes A group of chemicals, including formaldehyde and glutaraldehyde, all of which have in common the ability to cross-link protein molecules. Aldehydes are commonly used as fixatives in both light and electron microscopy.

Alternative Splicing The differential splicing of exons of the same primary RNA transcript to generate different mRNA molecules that code for different but related proteins.

Amino Acids A class of organic compounds with a carboxyl group and an amino group attached to a central carbon atom known as the α carbon. Twenty L α-amino acids serve as the monomeric building blocks of all the proteins.

Aminoacyl-tRNA Synonymous with activated tRNA. A tRNA molecule with its cognate amino acid attached to the 3' end.

Amphipathic The property of a molecule that contains both hydrophobic and hydrophilic parts. Examples of amphipathic molecules include phospholipids and bile salts.

Ankyrin A protein on the cytoplasmic side of the red blood cell membrane. It anchors or links band 3 to the fibrous molecule spectrin.

Anomeric Carbon Monosaccharides commonly mutarotate (i.e., form α and β hemiacetals.) This isomerism is so common in carbohydrates that they are given the special name *anomers*. The carbon atom at which the α and β isomers occur is called the *anomeric carbon*.

Antibiotics A diverse group of molecules that are purified from bacteria and plants and used medicinally to fight the invasion of a pathogen. They work through different mechanisms, ranging from inhibiting protein synthesis to inhibiting the assembly of cell wall components in pathogenic bacteria. Examples include penicillin and streptomycin.

Antibodies (Immunoglobulins) Bivalent molecules produced by B cells of the immune system. Antigen binds to antibodies on the surface of B cells, the cells are stimulated to divide, and the antibodies are shed. The antibodies then bind tightly to a cell or molecule, signifying that it should be removed. Antibodies are also useful for protein purification.

Anticodon tRNA sequence of three nucleotides that is complementary to an mRNA codon. Base pairing between the codon and the anticodon aligns the tRNA on the ribosomes and allows the addition of the correct amino acids in the right sequence.

Antigen Any substance that is capable of eliciting an immune response. Antigens are usually proteins or large carbohydrates.

Antiport A membrane carrier system that transports two ions or molecules in opposite directions. Often the driving force is one molecule that is moving down its electrochemical gradient; the other transported molecule is moving against its electrochemical gradient.

Apical Usually used to describe the "top" of a cell opposite the basal surface. For instance, kidney cells have a planar basal surface attached via a basal lamina to the underlying substrate. The apical surface, however, has many microvilli that increase the absorptive area of the cell.

Apoptosis Programmed cell death, mediated by death genes. Cells are characterized by apoptotic bodies, many membrane blebs, and a reduction in size. Proteases such as ICE are activated, and DNA is clipped into nonrandom units, which appear as a DNA ladder on DNA gels. It occurs during development, for example, during tadpole tail resorption.

Apotransferrin Also referred to as iron-free transferrin. Transferrin is a protein that transfers iron into cells through receptor-mediated endocytosis. It releases iron in the cell and is recycled to the cell membrane, still bound to its receptor.

ATP (Adenosine 5'-Triphosphate) The major energy currency of the cell. In eukaryotic cells, some ATP is produced through glycolysis and substrate-level phosphorylation, whereas most ATP is produced in the mitochondria. ATP consists of adenine, ribose, and three phosphate groups. The release of the terminal phosphate generates a large amount of energy.

Attenuators Loops of RNA formed by base pairing of sequences within the same RNA molecule. These inhibit further transcription of the gene downstream.

Autocrine Describing cells which release a hormone which feeds back and stimulates the receptors on the same cell.

Autophosphorylation The phosphorylation of a molecule by itself. Typically such a molecule will have a protein kinase domain within the molecule, a binding site for the ATP that donates the phosphate, and amino acids that are autophosphorylated. An example is the insulin receptor that is autophosphorylated on tyrosine residues immediately subsequent to insulin binding.

Autoradiography (Radioautography) A process used in microscopy, gel electrophoresis, and other techniques that allows an investigator to locate a radioactive isotope. Autoradiography uses a film emulsion (microscopy) or an X-ray film (gel electrophoresis) that is etched by the radioactive decay particles. Such a film is called an autoradiogram.

Axon A long process (up to a meter long) that typically is responsible for propagating the axon potential in a neuron without decrement in signal. The axon's structure is maintained by a combination of both intermediate filaments (neurofilaments) and microtubules.

Bacteriorhodopsin A pigmented protein found in the plasma membrane of *Halobacterium halobium*, a bacterium that lives in salty environments. It can pump protons out of its interior in response to light.

Band 3 A transmembrane protein of 100,000 molecular weight isolated from the red blood cell membrane. It is so named because it was originally detected as the third band on an SDS gel.

Basal Referring to the "bottom" of a polarized cell. The basal portion is usually planar and attached to the basal lamina, connected to underlying tissue. It is the opposite of the "top," or apical, portion of a polarized cell.

Basal Lamina (Basement Membrane) A collagenous sheet underlying many epithelial cells. It often includes Type IV collagen, fibronectin, and laminin. The basal lamina acts as a filter, a biological "glue" to which epithelial cells can adhere, and a promoter of differentiation for a number of cell types.

Bilayer Sheets Two sheets of molecules sandwiched together. Bilayer sheets are often used in reference to the phospholipid bilayer of cell membranes.

Bile Acids Amphipathic molecules found in the bile secreted by the liver. The primary bile acids in human bile are taurocholic acid and glycocholic acid, which aid in fat emulsification and digestion.

Bip (Binding Protein) A chaperone and resident endoplasmic reticulum protein. It is structurally related to the hsp70 proteins and recognizes incorrectly folded proteins. Bip may also help pull proteins into the endoplasmic reticulum.

Bivalent Possessing two binding sites, such as those that occur on an antibody.

Black Membrane An experimental membrane produced between two water reservoirs. Black membranes were used to understand the structure and function of biological membranes. The black nature is due to light interference patterns.

Blastomeres Cells produced by the cleavage of an egg subsequent to fertilization.

Blastopore An invagination that occurs during gastrulation of the embryo. The dorsal lip of the blastopore, located at the vegetal pole, contains Spemann's organizer—an important signaling center of the embryo.

Blastula An early stage of the embryo characterized by blastomeres surrounding a central lumen called a blastocoel.

Bright-Field Microscope A type of microscope used primarily to examine fixed (dead) and stained samples. It is the oldest of the light microscope systems.

cAMP (Cyclic AMP) A nucleotide that is generated from ATP upon stimulation of adenylate cyclase via a ligand-hormone interaction. cAMP was the first "second messenger" discovered and is used as a signaling molecule by most cells. It is broken down to 5'-AMP by phosphodiesterases.

Cancer The pathological condition whereby certain cells divide without control, compromising the health of an individual.

CAP (Catabolite activator protein) Also known as cyclic AMP—binding protein; acts as the positive control for the *lac* operon. It attracts RNA polymerase, thus facilitating the initiation of transcription.

Carcinogens Molecules or energy (e.g., radiation) that can convert cells from normal to transformed (cancerous). They can act either directly or indirectly on cells. In the latter case they can be converted from inactive carcinogens to active carcinogens in the liver.

Carcinoma The most common form of human cancer. It is characterized by unregulated division of epithelial cells.

Carrier Proteins Membrane proteins that transport molecules across membranes.

Catabolite Repressor Mediates the repression of transcription of many genes involved in sugar metabolism (including the ones in the lactose operon) in bacteria.

Catalysis The process of facilitating a chemical reaction by reducing the activation energy.

Catalyze To facilitate a chemical reaction by reducing its activation energy. This process does not affect the equilibrium of a reversible reaction.

Cdc2 (Cell division Cycle 2 or Cdk1) A cyclin-dependent protein that works with cyclin B as the mitosis-promoting factor in the eukaryotic cell division cycle.

Cdk (Cyclin Dependent Kinase) Any of a group of proteins involved in the cell cycle. They are active only when bound to a cyclin. The Cdks stimulate movement through various parts of the cell cycle through phosphorylation.

Cdk1 (Cell Division Cycle 2 or Cdc2) A cyclin-dependent protein that works with cyclin B as the mitosis-promoting factor in the eukaryotic cell division cycle.

Cell Cycle The life cycle of the cell. G_O is the differentiated, nondividing phase; G_1 is the first part of the cycle that requires protein synthesis; S is the phase of DNA synthesis; G_2 is between S and M; M is the mitotic phase. The typical eukaryotic cell cycle takes 24 hours to complete.

Cell Line A clone of cells usually derived from either an embryo or a tumor. A cell line is regarded as immortal because it will always continue to propagate so long as it is replenished with fresh nutrients.

Cell Strain A clone of cells usually derived from typical animal or plant tissue. It is a mortal group of cells because the cells will eventually reach cell crisis (i.e., die) once they have divided a certain number of times. The life expectancy of animal cell strains is dictated, in part, by the age of the subject from which the cells were isolated.

Centrifuge A machine that can create gravitational forces used to separate cells, cellular structures, and molecules according to their differential behaviors in a centrifugal field. The separation is achieved owing to differences in the mass, shape, and density of the structures being separated.

Chain Elongation The process of adding amino acids to a growing chain during protein synthesis.

Channel Formers Molecules that embed themselves in membranes and pass ions down their gradients. They are a type of ionophore.

Chaperonin A molecule that "chaperones" proteins, frequently across membranes. Chaperonins often use ATP and work by keeping proteins in an unfolded array. An example is the heat-shock protein hsp70.

Chemiosmotic Phosphorylation A theory originally proposed by Peter Mitchell which suggests that the buildup of both a pH and an electrical gradient across a membrane can be coupled to the phosphorylation of ATP from ADP.

Chiral Possessing "handedness." Asymmetric carbon atoms that have four different groups covalently attached are said to exhibit right-handed and left-handed orientations in space. The chiral nature (handedness) of the asymmetric carbon atom produces the l and d isomers.

Chiral Carbon A carbon atom that has four different groups covalently attached.

Chlorophyll The main pigment molecule family in most vascular plants. Chlorophylls have unstable porphyrin pigments containing magnesium surrounded by five rings. One ring has a long hydrocarbon tail that helps anchor the chlorophyll to the membrane.

Chloroplasts Double-membrane organelles in plant cells that contain chlorophyll. They are the sites of chemiosmotic photophosphorylation, which occurs in the thylakoid membranes, as well as carbon fixation, which occurs in the stroma.

Cholesterol An amphipathic lipid that contains the four-ring steroid arrangement. Cholesterol is often found in cell membranes, where it regulates fluidity. It is also the precursor for most other steroids and is shipped into the cell through receptor-mediated endocytosis.

Chromatid One of a pair of homologous chromosomes formed during the S phase and joined at the centromere. The chromosomes separate during mitosis, with one chromatid going to each of the two new daughter cells.

Chromatin Found in the nucleus of eukaryotic cells and consisting of DNA, histones, and nonhistone proteins. Euchromatin is usually less dense than heterochromatin. Experiments suggest that euchromatin represents the DNA that is being actively transcribed.

Clathrate Structure An ordered cage of water molecules formed around hydrophobic molecules in an aqueous medium.

Clathrin A triskelion molecule that can assemble into a clathrin "basket." This basket is the source of electron density in the coat of the coated pit seen during receptor-mediated endocytosis. Clathrin is also involved with other endo- and exocytotic processes.

Codon Sequence of three nucleotides in DNA or mRNA which specifies that a particular amino acid be added during protein synthesis.

Cofactor Small molecules or metal ions required in certain enzyme-catalyzed reactions.

Confocal Microscope A microscope developed in the 1980s that uses a laser for imaging single spots that it can store in a computer and display as a complete image. The confocal microscope increases the practical limit of resolution of the microscope, since much of the stray reflected image generated in a standard microscope is deflected because of the confocal pinhole in the confocal microscope.

Conformation The exact shape of a molecule in space. It is determined by the atoms and is important because a small change in protein conformation can, for example, alter the activity of the protein. Conformations can be determined through X-ray crystallography.

Connexon The functional unit of the gap or electrical junction. It consists of a pore surrounded by a ring of six proteins. Ions or small molecules travel through the pore and cause signaling and electrical changes in cells joined by gap junctions.

Constitutive Describing biological processes that are always turned on and thus can't be regulated by an external stimulus such as a hormone. It is the opposite of "regulated."

Constitutive Expression The constant expression seen in biological processes that are always turned on and thus can't be regulated by an external stimulus such as a hormone. It is the opposite of "regulated expression."

Covalent Bond An attractive chemical force generated by equal sharing of one or more electrons between two atoms.

Critical Concentration That concentration of either tubulin or actin monomers above which polymerization of the microtubule or actin filament occurs. When the concentration of the local monomers falls below the critical concentration, the microtubule or actin filament disassembles.

Cyanide A toxin that stops the electron transport system in mitochondria and thus kills cells and organisms by decreasing the amount of ATP available to run metabolic processes.

Cyclin A molecule that can combine with cyclin-dependent protein kinases to activate different stages in the cell cycle. Cyclins increase or decrease in amount according to the cell cycle stage.

Cytokinesis An integral part of mitosis during which two new daughter cells physically separate from each other through cytoplasmic division. The process is mediated by both microtubules and actin filaments. It is distinct from karyokinesis, the division of the nucleus.

Degenerate With reference to the genetic code, having more than one codon that specifies a certain amino acid.

Dephosphorylation Removal of phosphate from a molecule by hydrolysis. The process of protein dephosphorylation causes an alteration in activity, often through change in shape.

Depolarization A change to a less negative membrane potential. It is often used in reference to neurons. When their resting potential goes from −70 mV to 0 mV, for instance, the cell is considered depolarized. Depolarization of neurons often generates action potentials that can lead to the release of neurotransmitter.

Desmosomes Electron-dense cell membrane junctions often found between epithelial cells such as keratinocytes. They are found in tissues where tensile strength is important. Desmosomes often have keratin filaments that connect them to interior parts of the cell as well as indirectly to other desmosomes. A hemidesmosome connects cells to the basal lamina.

Determination A term often used in the study of development. It refers to the ability of a cell to commit to a specific pathway of differentiation.

Diacylglycerol (DAG) A molecule consisting of glycerol linked to two fatty acid chains. It is a lipid created by the splitting of inositol phospholipids in response to an extracellular stimulus. DAG then activates protein kinase C. Phorbol esters are chemical mimics of DAG and can cause cancer in animal cells.

Differentiation A process that occurs during development whereby a precursor or stem cell undergoes gene activation, resulting in a highly specialized cell.

Dipoles Small charge separations within the same molecule due to a large difference of electronegativity between two atoms that make up the molecule.

Disulfide Bond A covalent linkage between cysteine residues in two different proteins or in different parts of the same protein.

DNA Polymerases A family of enzymes that polymerize DNA. They use the template strand to form a new complementary strand. Deoxyribonucleotides are used as precursors and added one at a time in the proper base-pairing sequence. A pyrophosphate is released as a consequence.

Domains Areas of common function found in a number of proteins or other molecules (e.g., the protein kinase domain). The term is useful in comparing similar or homologous domains between molecules.

Double Bond A type of covalent bond formed by equal sharing of two pairs of electrons between two atoms.

Ectoderm The outer covering of an embryo or adult. It usually consists of epithelial cells that serve a barrier function. It is one of the three major tissue layers of the embryo: ectoderm, mesoderm, and endoderm.

EGF (Epidermal Growth Factor) A hormone that stimulates a receptor, often causing cells to divide. The EGF receptor has a tyrosine kinase that can autophosphorylate tyrosine residues.

Electron Transport Chain A linked group of intermediates or electron carriers that can sequentially pass electrons from higher- to lower-energy intermediates. In mitochondria and chloroplasts both a pH and membrane potential are created that can generate ATP through the ATP synthase.

Electronegative Attracting electrons. An electronegative atom attracts electrons toward itself (or, similarly, is reluctant to give up electrons). Some elements are more electronegative than others.

Electronegativity The arbitrary scale that measures the electronegative property of elements. In the scale Linus Pauling defined, fluorine has an electronegativity of 1, while all other elements have electronegativities less than 1.

ELISA (Enzyme-Linked Immunosorbent Assay) Assays that are often performed in 96-well plates using antibodies to detect the amount of antigen present in a solution. ELISAs are excellent quantitative assays for soluble proteins secreted or retained by cells.

Emulsify Render hydrophobic substances soluble in an aqueous solution by formation of micelles through the mediation of amphipathic molecules like bile salts and detergents.

Enantiomers Two nonsuperimposable structures that are mirror images of each other.

Endocytosis The invagination of a membrane, ultimately forming a vesicle containing cargo molecules. Receptor-mediated endocytosis can import, for instance, iron and cholesterol.

Endonuclease A family of enzymes that cleave DNA. Some, such as restriction enzymes, are found in bacteria and are very useful for molecular biology. Others that are endogenous to eukarotic cells are calcium-activated and cleave DNA during programmed cell death, or apoptosis.

Endosomes Closed vesicles that form subsequent to endocytosis. They deliver their cargo molecules to a particular destination such as a lysosome.

Endosymbiosis The mutual beneficial relationship generated by the fusion of a protoprokaryote and a protoeukaryote that resulted in the formation of the current eukaryotic cell containing mitochondria and chloroplasts.

Endothelial Cell A type of cell that surrounds blood vessels. These cells contain tight junctions between them, limiting fluid movement across the endothelial layer; however, there is exchange of materials such as oxygen, glucose, and carbon dioxide across the layer. Endothelial cells are usually attached to basal laminas.

Enzymes Biological molecules that act as catalysts by lowering the energy of activation of biological reactions. Most enzymes are proteins, and their specific shape is important to their activities.

Epithelial Relating to epithelium. The epithelium is a continuous sheet of cells that covers the outside and inside surfaces of most animals.

Erythrocyte (Red Blood Cell) A biconcave disk cell the size of a capillary. Mammalian erythrocytes are anucleate. Erythrocytes contain hemoglobin and are primarily responsible for carbon dioxide–oxygen exchange.

Erythrocyte Ghosts Erythrocytes that have been emptied of their cytoplasmic contents. This occurs through the cytoplasmic lysis of the erythrocyte. Erythrocyte ghosts have been particularly useful in studying membrane phenomena owing to their simplicity and the fact that erythrocytes have few contaminating organelle membranes.

Ester A molecule formed by the creation of a covalent bond between the carboxylic carbon of an acid and the oxygen of an alcohol.

Eukaryotic Referring to eukaryotes. A eukaryote is a class of organisms that includes plants, animals, fungi, yeast, and others that are characterized by a membrane-bound nucleus and a variety of membrane-bound organelles.

Exon Segment of a eukaryotic gene that when copied into an RNA transcript is spliced together with other exons to form mature mRNA.

Exonuclease An enzyme that can cleave a terminal nucleotide from a nucleic acid.

Extracellular Matrix A complex arrangement of many different types of molecules, such as collagen and laminin, that have been secreted by cells. The basal lamina is an example of an organized extracellular matrix. The extracellular matrix acts as a filter as well as a surface that encourages the differentiation of cells.

Facilitated Transport A process by which a molecule is passed down its gradient using a protein transporter. The rate of transport exceeds that predicted by diffusion. The process is saturable, and the transport is in the direction of the chemical gradient. No ATP is required for facilitated transport.

Fatty Acid A molecule that usually has a long hydrocarbon tail with a carboxyl group at one end. Fatty acids are metabolized for ATP production in mitochondria and are major components of phospholipids in plasma membranes.

Ferrotransferrin Also called iron-loaded transferrin. Transferrin is a protein that transfers iron into cells through receptor-mediated endocytosis. It releases iron in the cell and is then recycled to the cell membrane, still bound to its receptor.

Fertilization Fusion of a male and female gamete to form a diploid zygote.

Fibroblasts Animal cells that are able to secrete collagen. They are found in areas such as the dermis of the skin and are important to the maintenance of the extracellular matrix. They are commonly used in tissue culture studies.

Fluid Mosaic Model The most recent model of the cell membrane, stating that the cell membrane is composed of many types of proteins within a phospholipid bilayer. Many of these proteins are free to move in the lateral plane of the membrane. Freeze-fracture and "capping and patching" experiments gave rise to this model.

Fluorescence-Activated Cell Sorter A machine that can sort cells by recognizing fluorescent tags attached to the cells to be sorted.

Fluorescent Referring to a molecule's ability to absorb light of one wavelength within the visible spectrum and release light at a slightly longer (i.e., lower-energy) wavelength. Fluorescein is an example. It excites at 485 nm and emits at 535 nm. Fluorescent probes are integral parts of fluorescence immunocytochemistry.

FRAP (Fluorescence Recovery After Photobleaching) A method of measuring membrane fluidity by using fluorescently tagged molecules. A laser is used to bleach some of the fluorescently tagged molecules in the cell membrane, and the time necessary for other, non-bleached, fluorescently tagged molecules to fill in the bleached area is used to calculate membrane fluidity.

Free Energy A measure of the potential energy of a system that is a function of its enthalpy and entropy.

Functional Group An atom or a group of atoms characteristic of a class of organic compounds and exhibiting the properties of that class.

G Protein One of a very large family of GTP-binding proteins that mediate cell signaling. They are usually activated by a hormone-ligand interaction, which causes them to change their affinity from GDP to GTP.

Gap Junctions Cell junctions that mediate the exchange of small molecules. Gap junctions are composed of connexons, which are the conduits between adjacent cells through which molecules can pass. Neurons can electrically communicate via gap junctions.

Gastrula A stage in embryological development that occurs immediately subsequent to the blastula. Three tissue layers are formed and a primitive gut appears.

Gastrulation The process of forming a gastrula. A gastrula is a stage in embryological development that occurs immediately subsequent to the blastula. Three tissue layers are formed and a primitive gut appears.

GDP (Guanosine Diphosphate) A nucleotide that is often phosphorylated to GTP, guanosine 5'-triphosphate. GDP-GTP exchange is important in the functioning of G proteins in second-messenger systems.

Gel Filtration Chromatography One of several different protein separation techniques using a column. Gel filtration separates proteins on the basis of their size according to their ability to pass through biologically inert, porous beads. Smaller proteins that go through the beads have a longer path length and elute later than larger proteins.

Gene The entire sequence of DNA that is necessary for the synthesis of a functional polypeptide or an RNA molecule.

Genetic Code The set of nucleotide triplets (codons) that specify the amino acids in protein synthesis.

Genome The total amount of genetic information carried by a cell or an organism.

Glucose A six-carbon sugar that is one of the major energy-yielding molecules in cells. Glucose can be stored as glycogen in animal cells and as starch in plant cells. Glycolysis in animal cells breaks down glucose to pyruvate, yielding a net 2 ATP molecules per molecule of glucose.

Glycerol An organic alcohol with three carbons carrying three alcohol groups. This is a common alcohol found esterified with fatty acids in animal fat.

Glycocalyx The combined assemblage of carbohydrate groups present on the outside of cells.

Glycogen The storage form of glucose in animal cells. It is a polysaccharide (i.e., glucose polymer) that appears as electron-dense aggregates in cells.

Glycolipid A membrane lipid that has a small carbohydrate moiety attached to it.

Glycolysis A ubiquitous metabolic process that breaks down glucose, yielding ATP. In the presence of oxygen in eukaryotic cells, the result is 2 ATPs synthesized and 2 pyruvate molecules formed per molecule of glucose.

Glycoprotein Any protein that has an oligosaccharide attached to it. Secreted proteins and membrane proteins are commonly glycoproteins.

Glycosaminoglycans (GAGs) Very long and straight molecules found in the extracellular matrix. They are composed of repeating pairs of sugars, one of which is an amino sugar. Examples are hyaluronic acid and chondroitin sulfate.

Glycosidic Bond A covalent linkage formed as a result of a condensation reaction between two monosaccharides. Carbon 1 of one sugar reacts with the hydroxyl group of a second sugar, with the loss of a water molecule.

Glycosylation The process of adding a carbohydrate moiety to a protein. This process occurs in both the rough endoplasmic reticulum and the Golgi apparatus.

Golgi A multiple-compartment organelle in which newly synthesized proteins are modified before entry into the trans Golgi reticulum, where they are subsequently sorted. It was named after the researcher who identified it.

Gray Crescent A gray band appearing in the fertilized egg exactly opposite the site of sperm penetration. It is gray because of an alteration in the pigmentation pattern and is often the future site of the dorsal side of the embryo.

GTP (Guanosine 5'-Triphosphate) A nucleoside triphosphate that is involved in protein synthesis, microtubular stability, and cell signaling.

Heat-Shock Proteins A specific group of proteins expressed in response to elevated temperature or other stressful stimuli.

Hemidesmosomes Found between epithelial cells and the basal lamina and responsible for the attachment of these cells to the underlying substrate.

Histochemistry The use of specific microscopy stains to determine the molecular nature of cells and tissues.

Histones Highly conserved family of small basic proteins that associate with eukaryotic DNA in nucleosomes, which serve to package the DNA. The five main types are H1, H2A, H2B, H3, and H4.

HIV Human immunodeficiency virus, the virus believed to be the causative agent of acquired immune deficiency syndrome (AIDS).

Homo-oligomers In reference to a protein, a protein of identical multiple subunits.

^{3}H-Thymidine A radioactive compound also known as tritiated thymidine. It is an isotope that is commonly used to radioactively tag DNA often during DNA synthesis in the cell cycle.

Hybridization A process, integral to many techniques in molecular biology, that relies on the reassociation or reannealing of two complementary strands of DNA or DNA-RNA. It is a key part of Southern and Northern blots as well as *in situ* hybridization.

Hydrogen Bond A noncovalent interaction between an electronegative atom and a hydrogen atom covalently linked to another highly electronegative atom.

Hydrophilic Water-loving; capable of interacting effectively with water and able to dissolve in water.

Hydrophobic Can be translated as "water hating" and refers to nonpolar molecules that prefer a lipid rather than a water environment. Lipophilic is a synonym.

Hyperpolarization A term often used to describe changes in membrane potential in neurons when the resting potential moves farther from zero, as from –70 mV to –90 mV. It is the opposite of depolarization.

Hypotonic Often used to refer to a solution that has a lower solute concentration than a cell it surrounds. As a consequence, the water from the hypotonic solution diffuses into the cell to balance the difference in solute concentration, and the cell swells. Hypotonic solutions are useful for generating erythrocyte ghosts.

Immunocytochemistry The use of antibodies to identify antigens in cells employing microscopy techniques. Often antibodies are labeled with fluorescent probes so they can be tracked. Alternatively, in ultrastructural immunocytochemistry, instead of the fluorescent probe, an electron-deflecting gold particle is used.

in situ Means "in place." Thus *in situ* hybridization refers to hybridization in the intact cell, as opposed to hybridization of extracted nucleotides.

in situ Hybridization A technique that identifies specific gene or mRNA sequences within intact cells through the use of a tagged, complementary probe.

in vitro Means "under glass." It has two different meanings. It can refer to cells growing in culture, or alternatively, it can describe metabolic processes taking place in an isolated cell-free system, as might occur in the laboratory.

in vivo Refers to occurrences in an intact animal, plant, or other organism.

Inducer A molecule that is able to increase the expression of a gene or a set of genes (operon).

Inositol A cyclic molecule that is the hydrophilic part of inositol phospholipids.

Integral Membrane Protein (Intrinsic Membrane Protein) A cell membrane protein that interacts with the hydrophobic portion of the cell membrane to the extent that it can be extracted only using detergent solubilization. Plasma membrane receptors are examples of integral membrane proteins. Many integral membrane proteins traverse the entire membrane.

Interphase That part of the cell cycle that does not include mitosis. Thus, interphase consists of G_0, G_1, S, and G_2.

Intron Noncoding sequence of DNA present in the primary transcript of RNA and subsequently removed in the processing (splicing) of the RNA transcript.

Ion Carriers Molecules that carry ions from an area of higher concentration to an area of lower concentration. The ion carrier dinitrophenol is a lipid-soluble, weak acid that shuttles H^+.

Ion Exchange Chromatography One of several chromatographic techniques for separating proteins. Ion-exchange chromatography relies on protein binding to charged beads, and thus retarding of the protein movement through a column. For instance, negatively charged proteins will bind to positively charged DEAE beads.

Ionophores Any compound that can increase the permeability of a membrane to a particular ion. For instance, ionomycin can increase Ca^{2+} membrane permeability.

Karyokinesis The physical separation of the nucleus during mitosis that results in two daughter nuclei. It precedes cytokinesis.

KDEL Stands for "Lys-Asp-Gly-Leu." It is a sequence contained in endoplasmic reticulum–specific proteins that are retained in the rough endoplasmic reticulum and returned from the cis Golgi. Proteins containing the KDEL sequence are recognized by the KDEL receptor.

Kinetochore A multiple-layered structure found at the centromere of each chromosome during mitosis. Microtubules attach to the kinetochore; the movement of chromatids to opposite poles during mitosis (anaphase) relies on this attachment.

Krebs Cycle (Citric Acid Cycle; Tricarboxylic Acid Cycle) Consists of nine steps, occurring in the matrix of the mitochondria. Carbon dioxide, NADH, and $FADH_2$ are produced. Ultimately NADH and $FADH_2$ can be used to generate ATP through chemiosmotic phosphorylation.

Lagging Strand The DNA replication strand that consists of very short, discontinuous strands synthesized in the 5'-to-3' direction during DNA synthesis. These segments are called Okazaki fragments in honor of their discoverer. They are ultimately tied together by DNA ligase. The overall direction of synthesis of the lagging strand is in the 3'-to-5' direction.

Lamins A family of intermediate filaments that form the nuclear skeleton. They are associated with nuclear pores, and their phosphorylation is a key regulatory event during mitosis. Phosphorylation triggers nuclear dissolution; dephosphorylation is associated with nuclear reorganization.

Large Ribosomal Subunit 60S (in prokaryotes) or 70S (in eukaryotes) ribosomal subunit.

Leading Strand The continuous strand of new DNA that is synthesized in the 5'-to-3' direction during DNA synthesis. Unlike the lagging strands, no DNA ligase is necessary. The direction of synthesis is in the same direction as the replication (growing) fork.

Lectins A group of compounds, many of which are isolated from plants, that can bind tightly to proteins. Some, like ricin, are highly toxic. Others, however, are useful because they can be used either in situ or in affinity chromatography to purify proteins by binding to their carbohydrate groups.

Leukemia A classification of diseases often characterized by the overproduction of leukocytes, or white blood cells.

Leukocytes (White Blood Cells) A nucleated blood cell family that includes macrophages, neutrophils, and lymphocytes, all important in fighting disease in the immune response.

Ligand Any molecule that binds tightly to another macromolecule, as when a hormone (the ligand) binds to its receptor. A ligand-receptor complex often triggers a cell-signaling event.

Ligase An enzyme that links together the 3' end of one DNA strand with the 5' end of another to form one DNA strand.

Lipophilic (Hydrophobic) Means "lipid loving." Lipophilic molecules are nonpolar molecules that prefer to be in a lipid environment rather than a water environment.

Liposomes Pure, synthetic phospholipid bilayers useful for delivering hydrophilic molecules to the interior of cells through liposome-cell fusion.

Lysogeny The process by which an invading virus can splice its DNA into the genome of a recipient host cell. The viral genome remains dormant until a later time, at which the cell may enter the lytic cycle and new virions erupt from the cell.

Lysosomes Acidic, single-membrane organelles that contain enzymes designed to hydrolyze nearly all biological molecules. Lysosomal enzymes operate optimally at an acidic pH.

Lytic Describing the process in which viruses invade a host cell and cause the lysis of the cell through the production of new virus particles.

Macromolecule Any large molecule (usually made of many monomeric building blocks) with a molecular weight of more than a few thousand daltons.

Macrophages Large phagocytic cells involved in both primary and specific immune responses against substances foreign to the animal and human body.

Major Histocompatability Complex (MHC) A number of genes which encode the class I and II MCH molecules which bind peptides and present them to T-cell receptors. Different MHCs can result in tissue graft rejection.

Meiosis In eukaryotes a specialized process of cell division that results in haploid cells from a diploid cell. Cells double their DNA once, but there are two subsequent nuclear divisions, resulting in half the DNA present in the original cell. Meiosis generates gametes important to fertilization.

Mesoderm The middle layer of tissue produced during embryogenesis.

Metastasize To undergo metastasis. Metastasis is the movement of tumor or cancer cells from one place to another where another tumor develops as a result.

Metastatic Relating to metastasis. Metastasis is the movement of tumor or cancer cells from one place to another, where another tumor develops as a result.

Met-tRNA A tRNA molecule that carries a methionine attached to its 3′ end.

Micelles Concentrically arranged phospholipid or detergent molecules that spontaneously organize in a hydrophilic environment. The hydrophilic heads face the hydrophilic environment, whereas the hydrophobic heads are clustered together and face inward away from the hydrophilic environment.

Microscopy The process of using a microscope to examine the internal structure of cells and tissues.

Microtome A device used to slice thin sections of fixed or frozen tissue for observation in a microscope. A microtome for light microscopy can typically cut sections 10 to 15 μ thick—about the thickness of one cell.

Microtubules Long, cylindrical structures that make up one component of the cytoskeleton. They are composed of tubulin dimers arranged in 13 protofilaments. Microtubules have a distinct polarity, with (+) and (–) ends, and can polymerize or depolymerize according to the local concentration of tubulin dimers and other factors.

Missense Referring to a genetic mutation that alters a codon, causing it to specify an incorrect amino acid during protein synthesis.

Mitochondria Double-membrane organelles found in eukaryotic cells. They are self-replicating and can carry out protein synthesis. They are responsible for the energy demands of the cell through the production of ATP by oxidative phosphorylation.

Mitosis The physical separation of a dividing cell into two new daughter cells. Mitosis occurs after DNA synthesis and in mammalian cells lasts for only an hour. It is characterized by the dissolution of the nucleus, karyokinesis, and cytokinesis. Mitosis is subdivided into prophase, metaphase, anaphase, and telophase.

Mixed Micelles A concentrically arranged group of hydrophobic molecules of mixed variety. A typical mixed micelle is one that is generated by detergent solubilization of cells, resulting in a mixed micelle containing phospholipid and detergent molecules with their polar heads facing outward, their hydrophobic tails facing inward.

Modular In reference to a protein, describing one with multiple functions carried out by distinct regions. Functional domains of a modular protein, when attached to unrelated proteins, sometimes retain the original function.

Morphogenesis The change in form, usually of an embryo during development.

Morphogens Molecules secreted by a site in an embryo that can elicit morphogenesis or differentiation, often in a radial pattern originating from that site. It is thought that morphogens also provide positional clues to cells during development. Few morphogens have been chemically defined.

Motif In proteins, a sequence of amino acids that exhibits a particular three-dimensional structure that is usually associated with a specific function.

MPF (Mitosis-Promoting Factor; Maturation-Promoting Factor) A protein macromolecular complex that contains cyclin and a protein kinase that triggers mitosis. It was discovered simultaneously in both fertilized eggs and dividing somatic eukaryotic cells.

mRNA Messenger RNA, the molecule that carries the genetic instructions specifying the amino acid sequence of a protein during its synthesis.

MTOC (Microtubule-Organizing Center) A part of the cell that organizes and stabilizes microtubules. MTOCs act as nucleating centers for microtubules. Basal bodies are examples of MTOCs.

Muscle Cells Cells that are contractile in nature. Muscle cells are considered to be smooth, skeletal, or cardiac and can be recognized by the arrangement of their actin and myosin filaments.

Mutant In genetics, an organism that carries a permanent, transmissible change in its genome that can result in an altered phenotype.

Mutation A permanent, transmissible change in the nucleotide sequence of a gene that can lead to a change or a loss of function.

Myosin A type of protein that drives movement and uses ATP as an energy source. It is present in both muscle and nonmuscle cells. Myosin II is the "thick filaments" in muscle cells that slide over actin filaments, accomplishing contraction. Myosin I is found more often in nonmuscle cells and is commonly located near the plasma membrane.

NAD (Nicotinamide Adenine Dinucleotide) A coenzyme that is part of the oxidation reaction and accepts a hydride ion. NADH is produced in the citric acid cycle in mitochondria and is a critical electron carrier in the generation of ATP through oxidative phosphorylation.

Necrosis A form of cell death distinguished from programmed cell death, or apoptosis. Cell necrosis is a pathological event not involving genes and is characterized by leaky plasma membranes and a swelling cell.

Neuroblastomas Cancer cells of the nervous system. Neuroblastomas are common in children. Some neuroblastomas have been actively studied in culture.

Neuron Another name for a nerve cell. Neurons are electrically excitable cells that can pass action potentials from the cell body down an axon, causing the release of neurotransmitters at the synapse.

Nitric Oxide One of the most recently discovered second-messenger molecules. It is distinctly different from other second-messenger molecules in two ways. First, it is membrane soluble, so it can signal to adjacent cells. Second, it is a gas.

NMDA (*N*-Methyl-D Aspartate) A nonnatural amino acid that stimulates NMDA glutamate receptors.

NMDA Receptor: (N Methyl-D Aspartate) One type of glutamate receptor present on glutaminergic neurons. The NMDA receptor is typically blocked by Mg^{2+} under low stimulation but released under high-stimulation conditions. It permits the passage of Ca^{2+} into the cell, leading to a possible release of NO, which causes long-term potentiation.

Nondisjunction The process in which chromatids fail to separate during anaphase. As a result, one cell may receive an extra chromosome, whereas the other cell might be missing a chromosome. Nondisjunction can result in trisomy 21, or Down syndrome.

Nonsense In reference to genetic mutations, one that alters a codon, producing a stop codon. This leads to production of truncated polypeptides.

Notochord A mesodermal rod of cells just ventral to the neural crest in the developing vertebrate embryo. In higher-level vertebrates it forms the core around which other cells assemble to create the vertebrae.

NTRC (Nitrogen Utilization Regulator C) A protein important in the regulation of nitrogen utilization in bacteria. The phosphorylated form of NTRC interacts with RNA polymerase to alter its activity.

Nucleases A group of enzymes that can cleave DNA or RNA. Some nucleases are contained in lysosomes. Others isolated from bacteria have been indispensable tools for the molecular biologist.

Nucleophilic Nuclear-loving; referring to an atom, ion, or molecule that is attracted to positive charges.

Nucleotide The small building-block molecule of nucleic acids, composed of a purine or pyrimidine base attached to a phosphorylated pentose sugar that is either a ribose (in the case of RNA) or a deoxyribose (in the case of DNA).

Nucleus An organelle surrounded by a membrane envelope that is studded with pores. The nucleus contains DNA and proteins and is the site of RNA synthesis. Prokaryotes do not have an organized nucleus.

Okazaki Fragments Short stretches of DNA synthesized in the lagging strand of the DNA replication (growth) fork. Okazaki fragments are linked together by DNA ligase.

Oligomycin A drug that blocks the F_OF_1 complex (ATP synthase) in mitochondria and thus inhibits ATP synthesis.

Oncogenes Genes that, when activated, can transform cells from a normal phenotype to a cancer cell. Many oncogenes develop from proto-oncogenes, genes that govern normal cell growth activity.

Oncoprotein A protein produced by an oncogene.

Oocytes Egg cells; these are large, do not bind easily to a substrate, and often have many storage proteins present in the cytoplasm.

Operator A short segment of DNA that binds repressor molecules and alters the expression of a gene or a set of genes downstream. Found in bacterial and viral genomes.

Orbital Any one of the regions of space around the center of an atom where certain electrons are found with the highest probability.

Organic Referring to compounds that have carbon as a primary component. A misnomer founded in the initial belief that these compounds are unique to life.

Oxidative Phosphorylation A mechanism in mitochondria and bacteria that generates ATP by passing electrons from one intermediate to another, in the process creating both a pH gradient and often a transmembrane potential. Oxygen is necessary, and ATP is generated via an ATP synthase.

Patching and Capping The aggregation of fluorescently tagged antibodies that are associated with proteins on membranes of living cells. The aggregation appears as a cap or a patch in the fluorescence microscope and is due to the bivalent nature of antibodies. Patching and capping were critical in demonstrating the fluid nature of plasma membranes.

Peptide Bond A covalent bond that links two adjacent amino acids in a polypeptide: Formed by a condensation reaction between the α-amino group of one amino acid and the α-carboxylic group of the other.

Peptidyl-tRNA tRNA that lies in the P site of a ribosome during protein synthesis; one that has a peptide attached to its 3' end.

Peristalsis The rhythmic movement of the gut that propels food through the system.

Peroxisomes Small organelles that are bounded by a single membrane and often have a crystalline appearance. They possess catalase and degrade molecules such as fatty acids and amino acids by using hydrogen peroxide.

Phagocytosis The process by which large particles, including pathogenic microorganisms, are engulfed by certain eukaryotic cells, such as macrophages.

Phase Microscope A microscope system that uses differences in refractive indices in a cell or tissue specimen to generate contrast. No dyes are necessary. The phase microscope is most commonly used to examine living cells in culture.

Phosphatidylinositol Part of phosphatidylinositol 4,5-bisphosphate (PIP_2), which is part of the protein kinase C second-messenger signaling system. PIP_2, a membrane phospholipid found in the plasma membrane, is split by a phospholipase that was previously activated by a receptor-activated G protein.

Phosphodiester Bond A bond in which two hydroxyl groups form ester linkages with the same phosphate group. Adjacent nucleotides in DNA and RNA are linked by these bonds.

Phospholipid A class of lipid molecules, usually composed of two fatty acids esterified to glycerol phosphate.

The phosphate is also esterified with one of many amino alcohols. Found abundantly in cell membranes.

Photophosphorylation The process in plants and other organisms whereby light interacts with the chlorophylls and other pigment molecules, releasing electrons. These electrons pass through a photosystem housed in a membrane, in the process creating both a pH gradient and a transmembrane potential. This proton-motive force is then used to generate ATP.

Photorespiration A process that occurs in the chloroplast and uses oxygen. Ribulose 1,5-bisphosphate is converted to phosphoglycolate and then glycolate. The glycolate is then shipped to the peroxisome and the mitochondria, where carbon dioxide is released. It is considered a wasteful process because it uses oxygen and releases carbon dioxide.

Polar Molecule A molecule that carries a net charge or one that has asymmetric distribution of charges owing to large differences in electronegativity of the atoms that make up the molecule.

Polycistronic mRNA An mRNA molecule that can code for many proteins. Found in bacteria and viruses.

Post-translational Referring to the events that occur after a protein is synthesized, especially modifications to the protein.

Power Stroke A stage of muscle contraction in which myosin returns to its rigor position, pulling on actin in the process.

Primary Structure In proteins, the linear arrangement of amino acids and locations of the covalent bonds formed within the protein.

Prokaryotic Having to do with prokaryotes. Prokaryotes are simple organisms, including the eubacteria and the archaebacteria, that have no membrane-delimited nucleus or any other organelles.

Promoter The sequence of DNA that determines the site of initiation of transcription by RNA polymerase.

Propeptide A protein which often has an extra peptide at the amino or the carboxy end which is ultimately cleaved. Collagen is an example, because it is secreted as a tropocollagen propeptide. The terminal peptides are important for its final assembly in the extracellular environment.

Proteases Enzymes that cleave proteins at specific sites. There are several groups of proteases, classified according to the specific amino acids they recognize within the proteins to be hydrolyzed. Trypsin is an example of a protease.

Protein Kinase C A second-messenger signaling system found in many cells. Hormones activate a G protein, causing a plasma membrane–bound phospholipase to cleave phosphatidylinositol 4,5-bisphosphate into two signaling molecules. One of these activates protein Kinase C, which then phosphorylates substrate proteins.

Protein Separation A collection of processes including ion-exchange chromatography, gel filtration, and gel electrophoresis, all of which separate different species of proteins.

Proteins A class of biological molecules consisting of amino acids linked together by peptide bonds. They function as enzymes, provide structural support, and can be antibodies, hormones, and electron carriers.

Proteoglycans A group of glycoproteins that contain a central protein attached to one or more glycosaminoglycans. They are found in the extracellular matrix.

Proteosome A multiprotein complex of proteolytic enzymes that degrades unwanted cellular proteins.

Proto-oncogenes Genes that normally occur in eukaryotic cells and are involved in cell growth and division activities. Mutations or disregulation of proto-oncogenes can convert them into cancer-causing oncogenes.

Proton-Motive Force The energy contained in the pH differential and the transmembrane potential generated across membranes in chloroplasts, mitochondria, and other organisms by electron transport. The proton-motive force can be converted to ATP by proton flow through the ATP synthase.

Purine A basic compound found in nucleic acids. Composed of two fused heterocyclic rings.

Pyrimidine A basic compound found in nucleic acid. Composed of one heterocyclic ring.

Random Coils Stretches of amino acids found in proteins where the amino acids are arranged in space without any specific secondary structure.

Ras A guanine nucleotide–binding protein that occurs in signaling pathways in eukaryotic cells. It can be converted from the inactive Ras-GDP to the active Ras-GTP. Many oncogenes are related to unregulated Ras.

Receptor Tyrosine Kinase A type of receptor found in cell membranes of eukaryotic cells. Hormones bind to this receptor and cause an autophosphorylation of tyrosine residues in the cytoplasmic portion of the receptor. An example is the EGF receptor.

Receptor Any molecule, usually a protein, that interacts with a ligand and is activated in some manner. Receptors occur in plasma membranes, the cytoplasm, and the nucleus and are necessary in the hormone signaling process. Some work by activating second messengers, while others function by autophosphorylation.

Replication Fork (Growing Fork) In both prokaryotes and eukaryotes, the structure formed during DNA synthesis when the template strands separate and new complementary DNA strands are added. The leading strand is synthesized in a continuous manner, whereas the lagging strand is synthesized in short pieces called Okazaki fragments.

Repressor A protein encoded by a regulator gene. A repressor binds to an operator and represses the transcription of a gene or a set of genes.

RER (Rough Endoplasmic Reticulum) The major protein synthetic machinery of eukaryotic cells. Most membrane-bound and secreted proteins, but not soluble proteins, are made at this site.

Residue A single monomer of a large polymeric molecule, such as an amino acid in a protein.

Resonance The conceptual representation of the distribution of electrons of a chemical species among several different structural forms, used to explain higher-than-expected stability. Molecules like benzene rings can be depicted in two different ways by structures known as Kekulé structures. However, the real structure of the benzene is an intermediate state between the two Kekulé structures. Such a state is said to be due to resonance.

Restriction Enzymes (Restriction Nucleases; Endonucleases) A group of enzymes that can cleave double-stranded DNA at restriction sites. They are commonly found in bacteria, where they are thought to be useful in protection against invading viruses. Restriction enzymes are important tools in molecular biology.

Restriction Fragments Pieces of DNA that result from the action of restriction nucleases. Restriction nucleases are enzymes that can cleave double-stranded DNA at restriction sites. They are commonly found in bacteria, where they are thought to be useful in protection against invading viruses.

Restriction Nucleases (Restriction Enzymes; Endonucleases) A group of enzymes that can cleave double-stranded DNA at restriction sites. They are commonly found in bacteria, where they are thought to be useful in protection against invading viruses. Restriction nucleases are important tools in molecular biology.

Retinal Referring to the retina, the light-sensitive outpost of the brain in the back of the eye that consists of rods, cones, and other secondary neurons.

Retrograde Signal A signal sent in a direction opposite the accepted convention. For instance, postsynaptic neurons can release nitric oxide, and it can feed back to the presynaptic glutaminergic neuron in a retrograde manner to signal the presynaptic cell to release more neurotransmitter.

Retrovirus An RNA virus that replicates in eukaryotic cells by first making a DNA copy from its RNA using reverse transcriptase, which it encodes. This DNA template is then used to make mRNA for viral proteins as well as retroviral genomic RNA. HIV is an example of a retrovirus.

Reverse Transcriptase An enzyme that can make a DNA copy from an RNA template.

Ribosomes A multimeric complex consisting of many different rRNA species and several dozen proteins organized into a large and a small subunit. Ribosomes play an important role in protein synthesis and are the reason rough endoplasmic reticulum is "rough."

RNA Processing Modifications that are done to the primary transcript of RNA to make the functional RNA molecules.

Rough Endoplasmic Relating to the rough endoplasmic reticulum. The rough endoplasmic reticulum is the major protein synthetic machinery of eukaryotic cells. Most membrane-bound and secreted proteins, but not soluble proteins, are made at this site.

Rough Endoplasmic Reticulum (RER) The major protein synthetic machinery of eukaryotic cells. Most membrane-bound and secreted proteins, but not soluble proteins, are made at this site.

RUBISCO Stands for "ribulose 1,5-bisphosphate carboxylase." RUBISCO is the key enzyme of the Calvin cycle in the chloroplast stroma, where it fixes carbon dioxide, ultimately producing sucrose in the cytoplasm. RUBISCO has both carboxylase and oxygenase activities, the latter being wasteful.

Sarcoma A cancer of the connective tissue.

Sarcomere The functional unit of skeletal muscle. It extends from one Z disk to another Z disk and shortens during contraction. A sarcomere contains actin (thin filaments), myosin (thick filaments), and other contractile proteins.

Sarcoplasmic Reticulum A specialized organelle derived from the endoplasmic reticulum. It stores calcium and can be induced to release it upon nerve stimulation (depolarization). The calcium, in turn, elicits muscle contraction.

SDS (Sodium Dodecylsulfate) A detergent commonly used to solubilize cell membranes. It consists of a polar head group and a long, nonpolar hydrocarbon tail.

SDS Gel Electrophoresis (SDS-PAGE) A protein separation technique that uses an acrylamide gel, a power supply to generate a voltage difference across the gel, and, often, protein denaturing agents. Proteins are separated on the basis of molecular weight. This technique is both quantitative and qualitative.

SDS-PAGE (SDS Gel Electrophoresis) A protein separation technique that uses an acrylamide gel, a power supply to generate a voltage difference across the gel, and, often, protein denaturing agents. Proteins are separated on the basis of molecular weight. This technique is both quantitative and qualitative.

Second-Messenger One of a variety of molecules that act as intermediaries between hormone-receptor activation and ultimate changes in cell behavior. The first second-messenger discovered was cAMP, a signaling molecule used by most cells.

Secondary Structure In proteins, local foldings of the polypeptide chain into regular structures like α helices, β pleated sheets, and U-shaped turns and loops.

SH2 A domain on a protein that often binds to receptor tyrosine kinases once they have been activated (i.e., phosphorylated). These proteins then cause a change in cell behavior.

Signal-Anchor Sequence A sequence of amino acids at the amino end of a newly synthesized protein in the rough endoplasmic reticulum. This sequence is hydrophobic and, as such, anchors the protein to the membrane.

Signal Sequence (Signal Peptide) The first two dozen amino acids at the amino end of the protein that direct a newly synthesized protein to the rough endoplasmic reticulum. In the case of secreted proteins it is usually clipped off by the signal peptidase in the lumen of the rough endoplasmic reticulum. In other cases, however, it is retained.

Small Ribosomal Subunit 30S (in prokaryotes) and 40S (in eukaryotes) subunit of the ribosomes.

Southern Blot A technique that can detect specific DNA sequences. DNA is cut into pieces, separated on a gel, and then transferred to a paper blot. A labeled probe is then added that is complementary to a specific DNA sequence to be detected.

Spectrin A fibrous, tetrameric molecule that is a major part of the cytoskeleton in the red blood cell. Defects in spectrin result in misshapen red blood cells.

Spemann's Organizer A specialized area at the dorsal lip of the blastopore in developing amphibian embryos. It releases chemical messengers that determine the body axis.

Splicing The process of joining exons and removing introns of a primary RNA transcript to form the functional RNA.

Spot Desmosomes Single desmosome junctions that occur between epithelial cells. Desmosomes are characterized by electron-dense plaques with keratin filaments. They are important to tensile strength in the tissues containing them.

SRP (Signal Recognition Particle) A complex of one RNA molecule and six protein subunits that binds to the signal peptide of a nascent protein on free ribosomes. It temporarily halts protein synthesis and aids ribosome binding to the rough endoplasmic reticulum.

SRP Receptor The receptor on the endoplasmic reticulum that binds the SRP. The SRP is a complex of one RNA molecule and six protein subunits that binds to the signal peptide of a nascent protein on free ribosomes. It temporarily halts protein synthesis and aids ribosome binding to the rough endoplasmic reticulum.

START A critical checkpoint in eukaryotic cells. Passage through START commits the eukaryotic cell to cell division.

Stop Codon A codon on mRNA that specifies no amino acid and causes termination of protein synthesis.

Stop-Transfer and Anchor Sequence A hydrophobic domain within a newly synthesized protein that causes the protein to become firmly embedded in the rough endoplasmic reticulum membrane.

Substrate A molecule on which an enzyme acts, usually changing its activity.

Substrate-Level Phosphorylation The process in glycolysis whereby ADP is converted to ATP by a phosphorylated substrate.

Subunits Component parts of a large entity, often a macromolecule. For instance, a ribosome consists of one small and one large subunit.

Sucrose A sugar produced in the cytoplasm of plant cells as a consequence of carbon fixation in chloroplasts.

Symport A form of co-transport whereby two molecules are transported in the same direction across a membrane.

Synapsin A protein found in the axonal terminals of neurons. The phosphorylation of synapsin is thought to be important for the release of neurotransmitter.

Taq Polymerase A heat-resistant DNA polymerase isolated from a thermophilic bacterium that is a key element in PCR (polymerase chain reaction).

Termination Factor In protein synthesis, a protein factor that helps terminate protein synthesis once the ribosome arrives at the stop codon.

Terminator Codon Any codon in mRNA that does not specify an amino acid (UAA, UGA, and UAG). With the help of other protein factors, these codons cause termination of the translation of a polypeptide.

Tertiary Structure In proteins, the overall three-dimensional shape of the molecule, which is stabilized by multiple interactions between various amino acids.

Thymidine One of the five nucleosides that are precursors for DNA or RNA. Tritiated thymidine is the preferred radioactive isotope used for labeling DNA to track DNA synthesis.

Tight Junctions (Zonula Occludens) Specialized cell-cell membrane junctions commonly found in epithelial cells that line lumens filled with fluid. Tight junctions prevent fluid from leaking between two adjacent cells, so they form water-tight barriers.

Trans Golgi Network (Trans Golgi Reticulum) A specialized organelle functionally continuous with the Golgi. Proteins enter the trans Golgi network and are sorted according to their address tags. Some proteins are sorted to specialized vesicles destined for the lysosome; others are placed in vesicles that are secreted in a constitutive or regulated manner.

Trans Golgi Reticulum (Trans Golgi Network) A specialized organelle functionally continuous with the Golgi. Proteins enter the trans Golgi reticulum and are sorted according to their address tags. Some proteins are sorted to specialized vesicles destined for the lysosome; others are placed in vesicles that are secreted in a constitutive or regulated manner.

Transcript The initial RNA product formed by transcription of a gene.

Transcription The process of producing complementary RNA from a DNA template using RNA polymerase.

Transcription Factors Any protein other than RNA polymerase required to initiate or regulate transcription of genes.

Transfection The process by which a foreign piece of DNA is introduced into a recipient eukaryotic cell. Transfection is important for exploring gene function.

Transformation Most often used to refer to the conversion of a normal cell to a cancerous cell.

Transforming Growth Factor One of many hormones that can cause a change in behavior in a wide variety of cells. The TGF superfamily usually regulates the proliferation of most vertebrate cells.

Transitional Element (Transitional Vesicle) A small vesicle that transports proteins from the rough endoplasmic reticulum to the Golgi.

Transitional Vesicle (Transitional Element) A small vesicle that transports proteins from the rough endoplasmic reticulum to the Golgi.

Translation The process of protein synthesis on an mRNA template. Translation can occur on the rough endoplasmic reticulum or on free ribosomes in the cytoplasm.

Transmembrane Proteins Proteins that span cell membranes. Most hormone receptors are transmembrane proteins.

Tumor Suppressor A type of gene whose protein can actively inhibit the development of cancer cells from normal cells. $p53$ is a tumor suppressor gene whose dysfunction is responsible for over half of human cancers.

2D Gel Electrophoresis A protein separation technique that partitions proteins first by isoelectric focusing, then followed by SDS gel electrophoresis.

Ultramicrotome A device that uses either glass or diamond knives to cut thin sections of plastic-embedded tissue for subsequent examination in the transmission electron microscope.

Uniport A type of transmembrane transport system that moves one species of molecule in one direction.

van der Waal's Bond A weak noncovalent attraction due to a small, transient, asymmetric distribution of electrons around atoms.

Virions Individual virus particles.

Voltage-Gated Ion Channel A transmembrane channel found in excitable cells, such as neurons. A change in transmembrane electrical potential causes the channel to open, allowing ions to flow down their gradient. An action potential, for instance, is due to the activation of voltage-gated sodium channels.

Wobble The ability to base-pair in a nonstandard fashion between the third base of the codon on mRNA and the first base of the anticodon on tRNA. This enables some tRNAs to recognize more than one codon specifying a common amino acid.